SOFTWARE
SYSTEM
DEVELOPMENT
A GENTLE INTRODUCTION
FOURTH EDITION

WITHDRAWN

SOFTWARE SYSTEM DEVELOPMENT

A GENTLE INTRODUCTION
FOURTH EDITION

CAROL BRITTON & JILL DOAKE

The McGraw·Hill Companies

London	Boston	Burr Ridge, IL	Dubuque, IA	Madison, WI	New York
St. Louis	San Francisco	Bangkok	Bogotá	Caracas	Kuala Lumpur
Lisbon	Madrid	Mexico City	Milan	Montreal	New Delhi
Santiago	Seoul	Singapore	Sydney	Taipei	Toronto

Software Systems Development: A Gentle Introduction, Fourth Edition
Carol Britton and Jill Doake
ISBN-13 978-0-07-711103-8
ISBN-10 0-07-711103-6

 **Education**

Published by McGraw-Hill Education
Shoppenhangers Road
Maidenhead
Berkshire
SL6 2QL
Telephone: 44 (0) 1628 502 500
Fax: 44 (0) 1628 770 224
Website: www.mcgraw-hill.co.uk

British Library Cataloguing in Publication Data
A catalogue record for this book is available from the British Library

Library of Congress Cataloging in Publication Data
The Library of Congress data for this book has been applied for from the Library of Congress

Acquisitions Editor: Kirsty Reade
Development Editor: Karen Mosman
Marketing Manager: Alice Duijser
Production Editor: James Bishop

Text design by SCW
Cover design by Fielding Design
Typeset by MCS Publishing Services Ltd, Salisbury, Wiltshire
Printed and bound by CPI Group (UK) Ltd, Croydon, CR0 4YY

The McGraw-Hill Companies

BRIEF TABLE OF CONTENTS

DETAILED TABLE OF CONTENTS

ACKNOWLEDGEMENTS

We would like to thank Karen Mosman at McGraw-Hill for her support during this project. Many thanks also are due to the teaching teams in the Computer Science Departments at the University of Hertfordshire and Anglia Polytechnic University: Pam Hinton, Mariette Berkhout, Nathan Baddoo, Ros Crouch, Sandra Warren and Jacqui McCary, and all the students in both institutions who read the book, did the exercises and made (mostly repeatable) comments. Finally, our usual big thank you to Chris and Ol for putting up with us again during the trials and tribulations of writing the book.

We would also like to thank the following reviewers who commented on the previous edition:

Bjorne Rerup Schlichter, Aarhus School of Business, Denmark

Jim Moon, University of Glamorgan, UK

Thomas Chesney, Napier University, UK

Martyn Roberts, University of Portsmouth, UK

Pam Hinton, University of Hertfordshire, UK

PREFACE

This book is based on our own experience of teaching software system development. It is written for beginners who want to gain a sound grasp of the key concepts before moving on to more advanced topics. We do not assume any prior knowledge, and we try to explain everything in the simplest way possible. One of the most important aspects of the book is the graded exercises at the ends of chapters. Developing systems is a skill, and it is only by working through the exercises and practising the techniques that students will really learn to master them.

Our aim is to provide an understanding of the software system development process, and to give students the opportunity to become competent in some of the techniques involved in it. The book is based around the development of a small business system, from the initial identification of customer requirements, through the construction of models for the system, to the final implementation in a database package. Material covered and learning outcomes are listed at the start of each chapter. Where computer terminology is used, this is explained in the glossary at the back of the book. In the appendices there are two large case studies, designed to allow students to develop a system from the initial source documents through to implementation. Further models for these case studies (password-protected for use by lecturers) can be found on the website that accompanies this book, at http://www.mcgraw-hill.co.uk/books/britton.

New material in the fourth edition is mainly at the end of the book. In Chapter 8, which deals with implementation, there is a new section on SQL (Structured Query Language) and brief guidelines on designing websites. Chapter 10 has extra material on network charts, calculating the critical path, and legal and professional issues. In Chapter 11, on CASE tools and alternative approaches to development, there is new material on Agile Methods and open source development. In addition, the section on object-orientation has been rewritten and illustrated with examples from the Just a Line system to allow students to compare two very different approaches to software deveopment.

GUIDED TOUR

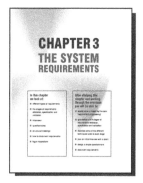

At the beginning of each chapter, there is an introduction that highlights what will be covered, and learning objectives pinpoint concepts you will learn about in the chapter.

Boxes, figures and illustrations bring the topics to life and help you to visualise software system development models.

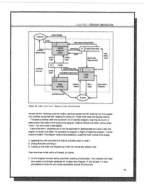

End of chapter summaries recap on the main points covered in each chapter, providing a useful revision tool.

Exercises and topics for discussion at the end of chapters provide you with questions to test your understanding of the concepts and apply them to real-life problems.

Further reading sections at the end of chapters point you to references in journals and books that can help you to research software system development further.

At the end of the book you will find answers to selected chapter exercises to allow you to monitor your progress. A glossary is also provided as a useful reference tool for your studies.

TECHNOLOGY TO ENHANCE LEARNING AND TEACHING

VISIT **WWW.MCGRAW-HILL.CO.UK/TEXTBOOKS/BRITTON** TODAY

Online Learning Centre (OLC)

Resources for students

After completing each chapter, log on to the supporting Online Learning Centre website. You will find chapter by chapter test questions which you can use to test your understanding of the topics covered in each chapter.

Resources for lecturers

The online learning centre also provides lecturers with a set of PowerPoint presentations which can be edited or adapted for use in lectures and seminars. Contact your representative for a password to access the material.

For lecturers: Primis Content Centre

If you need to supplement your course with additional cases or content, create a personalised e-Book for your students. Visit www.primiscontentcenter.com or e-mail primis_euro@mcgraw-hill.com for more information.

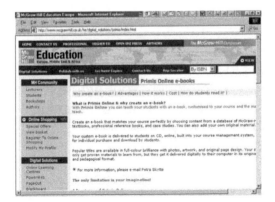

Study skills

We publish guides to help you study, research, pass exams and write essays, all the way through your university studies.

Visit **www.openup.co.uk/ss/** to see the full selection and get £2 discount by entering promotional code **study** when buying online!

Computing skills

If you'd like to brush up on your computing skills, we have a range of titles covering MS Office applications such as Word, Excel, PowerPoint, Access and more.

Get a £2 discount on these titles by entering the promotional code **app** when ordering online at **www.mcgraw-hill.co.uk/app**.

CHAPTER 1
BACKGROUND
AND CASE STUDY

**In this chapter
we look at:**

- an overview of the system
  development process

- characteristics of a system

- an overview of the material in the
  book and on the website

- the case study that will be
  developed throughout the book

**After studying this
chapter and working
through the exercises
you will be able to:**

- describe the main features of a
  system

- understand how to use this book

Introduction

This chapter serves as a guide for the rest of the book. In it we give an overview of the system development process, from the **client's** first tentative statement of the problem to delivery of the software system. We include a general description of what is meant by the word 'system' and a specific definition of the way it is used in this book. A summary is given outlining the contents of each chapter in the book and what you can find on the book website. We then introduce the Just a Line case study, which is used for examples and exercises throughout the book.

1.1 Developing systems

System development is a gradual progression from the client's initial vague ideas about the problem, via a series of transitional stages to a completely formal statement, expressed in a programming language, which can be executed on a machine. A diagram of the whole process can be seen in Figure 1.1. At each stage the problem is expressed in an appropriate modelling technique, notation or language.

This shows the system development process as a series of descriptions, or specifications, of the problem, which are gradually refined until a description is arrived at that can be understood by a computer.

The original accounts of the problem, which come from the clients, are most likely to be in a natural language, such as English, with perhaps a few diagrams and sample forms to help explain the situation. The final description will be in a programming language, since this is the only kind of language that a computer understands. What we are interested in is the question of which languages or notations to use at all the intervening stages of development.

Traditionally, **system developers/designers**[1] have worked within the context of a **framework** or **method** that provides an agreed structure for the development process.

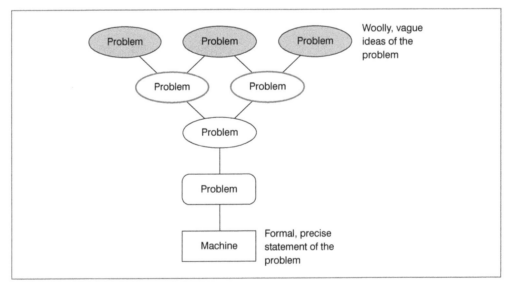

Figure 1.1 A view of the development process

---

[1] Words that appear in bold italic in the main text are defined in the glossary at the back of the book.

Normally, an organization will adopt a specific approach to developing a project, which it may refer to as the system or project life cycle, method or project plan. This framework provides a standard method of approach to the system developer's work, which is specified and documented before work starts. The same framework or method will generally be used for all projects developed within the organization.

However, it is increasingly being realized that different applications require different approaches. The approach used for a particular piece of software should be the one most suitable for the type of system being developed, the client, the developers and the techniques they have at their disposal. System developers are becoming aware that it is impossible for one single development method to prescribe how to tackle the great variety of tasks and situations encountered. In this book we concentrate on the traditional *structured approach* to developing systems, which is suitable for a small *information system*, such as that in the Just a Line case study introduced later in this chapter. In Chapter 11, we also briefly discuss the alternative approaches of *CASE*, *prototyping*, *Rapid Application Development (RAD)*, *Agile Methods*, *open source* development and *object-orientation*.

1.2 What is a system?

It is beyond the scope of this book to embark on a detailed discussion of system theory; the interested reader will find a fuller account in Myers and Kaposi (2004). However, it will be useful at this stage to have a working definition of the word *system*. If we look at some examples of systems:

- solar system
- digestive system
- public transport system
- central heating system
- computer system

we may arrive at a tentative definition – a system is a set of objects or elements that are viewed as a whole. On its own, this is inadequate for our purposes because we are concerned only with systems that are man-made, and therefore under human control, and that have a purpose – otherwise the system cannot be designed by a system developer. This rules out the solar system (no known purpose) and the digestive system (not under human control). An improved definition therefore would be: a system is a set of objects or elements that are viewed as a whole and designed to achieve a purpose.

Another essential feature in our view of systems is that the elements of a system have a relationship to one another; they work together in some way. A heap of stones, for example, although it may be man-made and have the purpose of marking the top of a hill, does not qualify as a system because the elements do not have a significant relationship to one another. If you take one stone from the pile, it does not matter much to the others. In a system removing one element would matter. If you remove the train service from the public transport system, it puts pressure on the other services – it affects them. If you remove the boiler from the central heating system, the system will not work. For our purposes, therefore, the definition of a system we need is: a system is an interrelated set of objects or elements that are viewed as a whole and designed to achieve a purpose.

We must add to this definition that a system has a boundary. The system in question lies inside the boundary; outside the **system boundary** is the **environment** with which the system interacts. Sometimes the boundary of a system is clear and obvious. If we view a person as a system the boundary is clear – one person is clearly separate from another and from the environment. In computer systems, however, it is usually hard to define the boundary; it is dictated by which elements we choose to think of as being within the system, and which as being part of the environment. The normal rule is that inside the system are things the system is designed to control; outside the system boundary are things the system interacts with, but is not designed to control. The boundary may be set because there are things over which we cannot have control – for example, in a central heating system the weather must be considered to be outside the boundary as we cannot control it. The boundary may also be set because we choose not to include certain elements. This choice may be dictated by the following factors.

- **Money constraints.** We may find that it will cost too much to computerize more than a limited set of system functions.
- **Time constraints.** The more functions we computerize the longer it will take.
- **Cost effectiveness.** Sometimes only limited benefits are gained from expensive computerization.

The environment is defined as being the surrounding conditions, outside the boundary, that affect the system and may be affected by it but not controlled by it. We might define the weather conditions as being part of the environment of a central heating system.

To summarize: a system is an interrelated set of objects or elements that are viewed as a whole and designed to achieve a purpose; it has a boundary within which it lies and outside of which is the environment. It is important to define the purpose or objectives of the system; different users will want different things from the system. From the outset, the system developer must be clear about the purpose of the system, as this is what will determine the design. The boundary in computer systems is often hard to define; the system designer and user together must decide what is to be included in the system. When we refer to a system in this book we base our understanding on the above description, but more specifically we interpret the objects or elements to be the **software**, **documentation**, method of operation, **hardware**, **users**, operators and **stakeholders**, which together make up the system.

The Just a Line case study that we use in this book is a particular type of software system – one that primarily stores, retrieves and manipulates data in a business environment. This type of system is known as an **information system**, or sometimes more specifically as a **transaction processing system**. The techniques that we describe in this book are all suitable for the development of a system such as Just a Line, and in many cases are also appropriate for a number of other types of software system development.

1.3 Contents of the book

In this book, we focus on the tools and techniques most widely used in a traditional structured approach to developing a software system. Chapter 2 outlines the main stages in the traditional system **life cycle** that underpins this approach, and relates the life cycle to the concept of a development method. Chapter 3 discusses the three stages of requirements engineering: elicitation, specification and validation. In Chapters 4, 5, 6 and 7 traditional modelling techniques are introduced and illustrated with examples from the Just a Line case study. Once modelling has been completed, implementation issues must be considered and many new factors become relevant; these are discussed in Chapter 8. Chapter 9 covers testing and what happens after delivery of the system. Chapter 10 provides an introduction to simple project management techniques, and Chapter 11 looks briefly at alternative development approaches: automated CASE tools, prototyping, Rapid Application Development, Agile Methods, open source development and object-orientation. There are two appendices, each containing introductory material and source documents for a case study. Each of these will provide opportunities for students to apply the techniques learnt in a coherent and challenging context, and allow them to develop a case study from initial stages through to implementation. Lecture slides are available on the book's accompanying website, at http://www.mcgraw-hill.co.uk/books/britton, password-protected for lecturers, together with multiple-choice exercises and answers.

You will get most benefit from this book if you read it through from start to finish and attempt at least some of the exercises in each chapter. Learning to develop software systems is like learning to swim: you can read and learn by heart all about swimming techniques, but if you don't actually practise, you will never learn how to swim. Working through the exercises in this book will give you essential practice in techniques for developing software systems. For those readers who are interested in particular topics, the book can be used as a reference manual, since each chapter can be read in isolation. As the same case study is used throughout, it is a good idea to read the introductory material on the Just a Line system below and the interview in Chapter 3. No technical knowledge is assumed; where computer terminology is used, this is explained in the glossary at the back of the book. As the book is written for newcomers to software system development, we aim to present an overall view of the subject. Some topics are discussed at an introductory level only, with suggestions for further reading at the end of each chapter. A full bibliography can be found at the back of the book. All of the chapters in the book have exercises, and answers to the majority of these are provided.

1.4 Introduction to the Just a Line case study

In the autumn of 1996 Harry Preston went up to university to study psychology. His three years at university were very busy ones, but spent mainly on the football pitch and socializing. While these activities were certainly character building, they did little to enhance his store of knowledge of psychology, the sum total of which lamentably failed to impress the examiners in his final-year exams. He graduated in 1999, scraping a third-class degree, which worried him not a bit, though it caused his mother considerable social discomfort.

Harry's time at university was not entirely wasted though. One evening at a party he met Sue, a student at a local teacher training college and, in the summer of 1999, they set up home together in a tiny and very run-down terrace house in a small village. The only cloud on the horizon was that employers did not seem to be particularly impressed by Harry's work record – or lack of it. After six months of Harry's fruitless searching for a job, Sue realized that drastic action was needed. She gave up her teaching post at the local school, withdrew all their savings from the building society and set up their own company, Just a Line, specializing in designing personalized cards and selling them by mail order.

It is now six years on and the venture has proved to be a great success. Just a Line cards range from 'Classic' (plain white with the customer's address printed discreetly across the top) to 'The Real You', which are personalized card and envelope sets with a design from the Just a Line range. Card designs are selected from the company list, which includes such items as old masters, botanical studies and famous quotations. Sue has the organizing ability to run the company and Harry has proved to have a distinct flair for picking popular lines – his 'Endangered Flora of the Hedgerows' range and his off-beat cartoon characters proved real winners. Harry and Sue moved two years ago into another run-down cottage, but larger this time and with two outhouses in the garden. These they have turned into a store-cum-shop with a view to developing the local side of their operations.

However, mail order remains the bulk of their business. At present, most of this is done by phone, which is time-consuming and inefficient, especially if the shop happens to be busy. The Prestons have a list of card types and prices, which is updated regularly, but have not yet got round to organizing a proper mailing list of regular and potential customers. The card list is given to anyone who asks for a copy and is displayed in strategic positions locally, such as libraries, supermarkets and village halls. The company advertises weekly in the local paper, and in the national press two or three times before Christmas. This has proved increasingly worthwhile, with the result that large retailers are beginning to show an interest in selling a selection of the cards.

While delighted with this development, the Prestons realize that a move into this type of market will have a considerable effect on the way they run the company. Their present rather casual methods of ordering, stock control and accounting will need to be tightened up, and a more professional approach will have to be taken overall.

Since they are keen to make the most of the new marketing opportunities, Harry and Sue decide that the necessary reorganization will be greatly helped by introducing a computer system into at least part of the organization. They consult D&B Software, a local firm of system developers that specializes in the computerization of small businesses, and it is agreed that one of its staff members will investigate the possibility of developing a computer system for Just a Line.

The rest of this book uses the Just a Line case study to illustrate the work of a software system developer.

Exercises and topics for discussion
What can you remember?

You will find all the answers in the chapter
a) What is the definition of a system?
b) What dictates where the system boundary lies?
c) What is the environment of a system?

1.1 For each of the following systems, identify the purpose, the elements, the boundary and the environment:
  ■ an electric shower
  ■ a car
  ■ a school library.

1.2 There are many different models of the system life cycle. Look at some books on system development and compare the different versions of the life cycle that you find in them. Make a short list of the most common life cycle models.

1.3 Imagine that you have been asked to design an information system for your local area. Discuss the factors that will influence your choice of boundary for the system.

1.4 Discuss in a group whether each of the following is a system according to the definition given in this chapter:

- an egg
- a television
- a business organization
- a painting
- a computer.

For each of the above that you consider to be a system, identify where the system boundary lies.

References and further reading

Myers, M. and Kaposi, A. (2004) *A First Systems Book: Technology and Management*, 2nd edn, Imperial College Press, London.

Skidmore, S. and Eva, M. (2004) *Introducing Systems Development*, Palgrave Macmillan.

CHAPTER 2
LIFE CYCLES
AND DEVELOPMENT
METHODS

In this chapter we look at:

■ the need for a system life cycle

■ deliverables

■ the stages of a typical system life cycle

■ system development methods

After studying this chapter and working through the exercises you will be able to:

■ explain the role of the system life cycle

■ give details of the stages of a typical system life cycle

■ describe what a deliverable is in the context of system development

■ explain the difference between a life cycle and a development method

Introduction

Traditionally, the development of a system is divided into several main stages. The progression of a system through these stages is known as the system *life cycle*. In this chapter we discuss the reasons for splitting up the system development in this way, and the general nature and content of the stages.

There is no single generally accepted life cycle. Various development methods have evolved that precisely define the stages into which the development process should be split and the exact sequence of tasks to be performed at each stage. A development method gives a recipe for the development of systems. We introduce the concept of a development method and discuss the advantages and disadvantages of using one.

2.1 The system life cycle

The rapid increase in the power, speed and capacity of computers, and the demands of clients and the market-place have encouraged software developers to attempt to develop ever more ambitious systems. First attempts, in the 1960s and early 1970s, to develop large and complex systems were discouraging: typically, systems were delivered years late, over budget, were unreliable, difficult to maintain and did not do what was required. These problems were so prevalent that this time came to be known as the *software crisis*. It was realized that solving big problems needed new methods – not scaled-up versions of techniques used for solving small problems. The system life cycle (Figure 2.1) was an attempt to establish a structured approach to analysing, designing and building software systems; developers started to introduce rigour into their method of producing software. They aimed to be able to deliver their product with the same stamp of reliability as more established professions, such as civil engineering or architecture, could deliver theirs.

The system life cycle divided the development of a system into stages. It specified the general nature of the activities involved at each stage, the sequence in which these activities should be ordered and the output or *deliverables* from each stage. There were several advantages to this approach to system development. The activities involved at each stage were defined, documented and agreed. This helped when training new staff, and among established staff it meant that a consistent approach to system development was achieved. Communication between teams of system developers was also improved by adopting an agreed approach. For project managers, the advantage was that each stage could be used as a milestone. Managers could put a date to that milestone and use it to monitor the

```
Problem definition

Feasibility study

Requirements engineering

Design

Implementation

Maintenance
```

Figure 2.1 Stages in a typical system life cycle

development of the project. Having the activities involved at each stage specified beforehand brought tremendous advantages in terms of estimating the timescale for the project, and costing and controlling the system development (see Chapter 10).

Computer science is a relatively young discipline, which is still evolving. It has never agreed on a single right way to develop a system and, given the enormous diversity in the types of system it tackles, it probably never will. However, most structured system development approaches do partition the development process into a more or less agreed sequence of stages for the development of information systems. The client's **requirements** are investigated, expressed and agreed in logical terms before decisions about the physical design are made. This means that the system is initially designed in terms of what it must do (for example, record orders and keep track of changes of address) before deciding how this logical design should be physically constructed (for example, by using a **network** of PCs and a database package). At the logical stage, the **design** is deliberately expressed in non-technical terms so that it can be understood and checked by the client. Once the logical design is agreed, the physical design can be tackled; at this stage the system developer proposes the hardware and software that will meet the client's requirements. When this design is agreed, the system can be implemented.

As we said above, there is no one definitive system life cycle; examples of life cycle models that map onto the development stages shown in Figure 2.1 include the Waterfall, the Spiral and the V-Model. Other life cycle models are based on different approaches to development, such as incremental development or **prototyping** (see Chapter 11).

The stages in a typical life cycle are shown in Figure 2.1. The output or product of each stage in the life cycle is known as its deliverable. In the early stages, the deliverables will be reports or documents describing first the existing and then the proposed new system. At the end of each stage in the life cycle there is normally a review meeting with the client to examine the deliverables of that stage for correctness. Deliverables produced at the end of one stage normally serve as working documents for the next stage until, in the final stages, the deliverable is the system itself. Deliverables also form part of the documentation of the system.

The content of each stage – the steps and activities involved – may vary from one practitioner to the next. If a standard development method (see below) is used, precise details are given as to what should be done and what techniques should be used. A rough guide to the nature of each stage in the life cycle is outlined below.

Problem definition

The problem definition provides an initial description of the **problem domain** by means of a written statement of the client's current problems and the objectives of the new system. It is normally produced after initial meetings with the client, and it documents the system developer's understanding of the situation at that stage. The problem definition must be agreed with the client before progressing to the next stage. It provides a firm foundation for the rest of the project, ensuring that the right problem is being tackled. It normally takes the form of a report, divided into sections. Typically, the following points will be outlined:

- the problems, as stated by the client and interpreted by the developer
- the objectives of the new system
- the scope and size of the project – which areas are to be considered, who will be involved

- preliminary ideas, from both the client and the developer, on how the system might be developed
- recommended action for the next stage in the development of the system.

Figure 2.2 shows an initial problem definition for the Just a Line system. The problem definition is the first stage in constructing the requirements specification document, which records all problems and requirements that are mentioned by clients in interviews, or are subsequently discovered during *analysis* of the system.

PROBLEM DEFINITION – *Just a Line*

Problems

- Taking orders by telephone: difficult, time consuming and error prone, as customers normally do not have a list of card designs or prices to hand. This will get worse as business increases
- Lack of effective marketing strategy
- Only one supplier (may let the company down)
- Haphazard stock control

Objectives

- To ensure that customer orders are processed efficiently
- To ensure that ordering is made easier for customers
- To improve company advertising and marketing
- To improve stock control: provide accurate information about current stocks, current stock requirements, identification of fast-/slow-moving lines, and to ensure that at all times there are adequate supplies of all lines in stock
- To facilitate future expansion of the business

Scope

The project will encompass the following areas of the business:
- order processing
- sales organization and administration
- marketing
- invoicing, including dealing with customer credit accounts
- stock control, including working with more than one supplier
The project will not encompass general accounting, payroll, personnel

Preliminary ideas

Introduce a computer system to:
- improve customer ordering procedures
- keep track of sales and stock
- compile a mailing list and circulate regularly with latest designs, price list and order form
- handle customer payments, including credit accounts

Recommended action

- Investigate and produce recommendations for the introduction of computer-based systems for the areas highlighted in this report
- Research the costs and benefits of each proposed system
- Investigate the cost and potential of using a website to display the latest card designs
- Investigate the feasibility of allowing customers to order over the Internet

---

Figure 2.2 Problem definition for Just a Line

Feasibility study

The *feasibility study* investigates whether there is a practical solution to the problem outlined in the initial problem definition. At this stage there has been very little financial investment in the new project. The feasibility study is a precautionary survey, its purpose being to do just enough preliminary work to establish whether the problem is one that can appropriately be tackled by a system development team. If this proves not to be the case, then the project can be abandoned at this stage before any great commitment of funds has been made.

In particular, the feasibility study examines the technical, economic and organizational feasibility of the project:

■ Can it be done?
■ Can we afford it?
■ Will the proposed new system fit in with existing procedures?

The feasibility study will determine the criteria for a successful system, and propose and evaluate several alternative solutions. The economic feasibility is of paramount importance and

---

Feasibility Study Report

1. Introduction. Project history and background, reference to any preliminary work on the project, agreed terms of reference, problem definition.

2. Definition of boundaries and scope of project. The existing system is described, using such techniques as data flow diagrams and E-R models (see Chapters 4, 5 and 6). The boundaries and scope of the investigation are clearly identified. Only high-level overviews of the system are used at this stage; the models will be fleshed out with detail at the next stage.

3. Requirements. The new system usually has to do everything the old one does plus solve existing problems and meet new client requirements. All requirements, those carried forward from the existing system and known new requirements, should be identified. Any constraints should also be listed – for example, that the new system must run on existing hardware, interface with existing systems or use specific software.

4. Alternative solutions considered. Usually several possible solutions, several ways of meeting the requirements specified, are outlined in the feasibility study report. The alternatives proposed may have different hardware, software or automation boundaries. They will therefore have different costs, benefits and development timescales. The technical, economic and operational feasibility of each solution will be evaluated. A rough implementation schedule for each solution will be included.

5. Recommendations. Normally the analyst presenting the report will be expected to recommend one of the proposed solutions. Material will be presented to support this recommendation; normally a detailed cost–benefit analysis will be the most important part of the supporting material.

6. Project plan. A fairly detailed development schedule should be produced for the recommended solution. Estimated costs for each stage should be included.

7. Conclusions and recommendations. A clear, concise summary of the report, and the conclusions and recommendations will be given. This may be the only part of the report read by a busy manager, so it should summarize all important issues including main requirements, alternatives considered and rejected, the recommended solution, costs and timescales.

Figure 2.3 Structure for a feasibility study report

a careful cost–benefit analysis is done for each of the proposed solutions; financial benefits must demonstrably outweigh the costs. Alternative solutions may be assessed in terms of initial financial outlay only, the time it takes to recoup the investment or the long-term profitability. Consideration will also be given to what will happen if no new system is developed. Other criteria for success should be specifically listed and should include tangible results – for example, 'all orders must be processed within 24 hours' or 'increase by 75 per cent the number of customers using an order form'.

The feasibility study is effectively a high-level superficial run-through of the rest of the system life cycle, and will involve doing some preliminary work on subsequent stages, possibly even some physical design of the system. It will certainly involve more exploration of the client requirements and will suggest some alternative solutions with the developer's comments and recommendations.

At the end of this stage, a feasibility study report is presented by the system developer to the client and a decision is made whether or not to proceed. The feasibility study report will vary according to the client organization and the system, but will usually contain sections on scope, client objectives, performance requirements, interfacing systems, impact on the organization and other systems, costs, benefits, risks and the consequences of not developing the system. Figure 2.3 shows one possible structure for a feasibility study report.

This used to be a crucial stage in all systems. Feasibility studies often involved a great deal of work and could last several months, with a very real chance of a recommendation at the end of this stage not to proceed with the development of the system. However, as experience of producing systems has accumulated, and with the advent of prototyping and powerful rapid development tools such as *CASE* and high-level programming languages, developers are more confident about what can or cannot realistically be achieved and so the feasibility study is neither as time-consuming nor as significant as it used to be.

Requirements engineering

The process of discovering and agreeing with the client exactly what the problems are and what the new system is to do is known as *requirements engineering*. In many ways this is the most crucial stage of developing a system: if the developer has got the wrong idea of what the client wants, then all subsequent work on the system is a complete waste of time. Chapter 3 describes some current requirements engineering techniques and discusses the principal issues relating to this topic.

Current physical model

Facts will rarely emerge in a neatly ordered fashion. It is much more usual for system developers to discover a mass of detailed, incomprehensible and probably conflicting information, all of which has to be sorted out and documented in a way that will help to organize the material. This must then be discussed with the clients to check that the system developer has correctly understood the problems and requirements, fill in any gaps and resolve any apparent conflicts. The modelling techniques discussed in Chapters 4, 5, 6 and 7 are designed to help the system developer do this. It may be useful to begin by documenting how the system functions currently. In the Just a Line system this will include such details as:

- orders come in by telephone
- if Sue wants to check she has enough cards in stock to fill an order, she looks in the store
- orders are completed in triplicate and filed in the three-drawer filing cabinet.

Sometimes a system developer will not attempt to document these physical details. Whether or not it is a useful thing to do will depend on the circumstances. If the system developer is tackling a complicated system with many existing interrelated procedures, or if there are many different users with many different and apparently conflicting versions of how the existing system works, or if the developer feels it is particularly important to get right the detailed workings of the existing system, then it is sensible to document such details. Often, the users will respond well to this type of model of the system because it depicts the system as it currently works, showing physical details they recognize and can relate to.

This model is referred to as the current physical model. Producing a current physical model is extra work and, moreover, some developers feel that if too much attention is paid to how the system currently functions, the new design will be constrained by this view of the system – the developer will be accustomed to seeing the system in one way and will be unable to come up with a radically new and more appropriate design.

Current logical model

From the detail of how the existing system works, the developer must extract exactly what the existing system does, as the new system usually must incorporate most (or all) of these features, as well as solving known problems and meeting additional client requirements. This second model of the system is known as the current logical model; it confines itself to documenting basic system events – for example, that orders come in and are recorded – and omits the physical detail about how the order comes in (by telephone) or how it is recorded (stored in a filing cabinet). The logical events (orders coming in and being recorded) will almost certainly be perpetuated in the new system, which is why the developer needs to record them in the current logical model. The physical details (certainly the fact that orders are recorded in a filing cabinet) we would expect to change if a computerized system is introduced. Therefore, these can safely be dropped from the logical model.

Sometimes a developer will start with the current logical model and omit the current physical model. Sometimes there is no existing system to model and the developer must start at the next stage – the required logical model.

Required logical model

The requirements engineering stage moves from the logic of the current system – what the existing system does – to the logic of the required or new system. It aims to specify what must be done to solve the problems and meet the requirements specified in the problem definition and the feasibility study. However, at this stage, analysis must be done without making decisions about how the new system is to function. The design at this stage is *implementation independent* and can therefore be implemented in several different ways. This is discussed further in Chapter 8. Experience has shown that if implementation decisions are made too soon, the design of the new system can be unnecessarily constrained by the limitations of the hardware or software selected. For example, if the developer decides at an early stage that the system will be implemented using a commercial database package, this may preclude the opportunity to use a file design that would have been more suitable. If the developer is committed from the start to a certain type of computer, this may mean that it is impossible to use a particular item of software that would have been ideal for the system. Sometimes there is no choice in the matter, it may be necessary to use existing client hardware, but if there is a free choice, decisions about implementation issues should not be made at this stage.

The deliverable from this stage is the specification of requirements, a logical model of the required system that states what the system is to do, but that says little or nothing about how the system is to be implemented. The model usually includes a data model and data flow diagrams of the required system, a supporting data dictionary and process definitions. It may also include preliminary discussions on sizing, performance requirements, security and the user interface.

Design

Once requirements engineering is complete, the next step is to determine how, in general, the problem is to be solved. The deliverables of the system design stage will be outlines of several different technical solutions that will meet the requirements specified in the previous stage. These alternative solutions usually include:

- a minimum-cost solution, which does the job and no more
- a medium-cost solution, which does the job well and is convenient for the user; it will probably have additional features the client did not request but which the system developer knows from experience will be needed
- a high-cost solution – everything the client could ever need, but at a price.

Different solutions may have features that differ in the following ways.

- **System boundaries.** The proposed systems may affect, though not necessarily computerize, different parts of the organization's functions. In Just a Line, one solution might include the accounting functions while another leaves them unchanged.
- **Automation boundaries.** One proposed solution might leave some functions to operate manually, while another computerizes them. In Just a Line, a supplier purchase order could be automatically generated when stocks fall low. Alternatively, the new system could simply generate, on request, a weekly list of items that are low in stock, or supplier ordering could remain completely manual.
- **Hardware.** One solution might propose the use of a large central computer with several terminals and laser printers. A cheaper solution might recommend the use of a network of personal computers (PCs) and less sophisticated printers.
- **Software.** The system may be implemented in different ways; it may, for example, be coded from scratch using a programming language, or it may be based on a pre-existing software package.
- **Design strategies.** The system developer may propose the development of a system using a traditional life cycle approach or using prototyping (see Chapter 11).
- **User interface.** The design of the user interface will be determined by the type of person using the system. Someone who is not used to computers and who uses the system only occasionally will need more help than someone who has been trained to use computers and sits at a screen all day. User interface design is a crucial part of all system development, since a system that users find unfriendly and hard to use will be rejected and must count as a failure.
- **Cost and time.** All the factors discussed above will affect the amount of money and the time it will take to develop the system.

Once one of the proposed solutions has been chosen, the new system must be specified in detail. This involves converting the implementation-independent specification from the requirements engineering stage into a design that includes specific hardware and software. This is often referred to as the technical design specification and may include:

- program design and specification
- specification of the user interface
- specification of the layout of reports and other system outputs
- file and record specifications
- hardware specifications, including costs
- implementation schedule.

Implementation

During the *implementation* stage the system is physically built: the program code is written and tested, and supporting documentation is produced. The deliverables from this stage of the life cycle include:

- program listings, test plans and supporting documentation
- details of the hardware on which the system will run
- manual of operating procedures
- manual of clerical procedures
- user manual.

The system must then be installed at the client's site on their equipment, and the changeover from the old to the new system supervised. This will involve training users and ensuring that

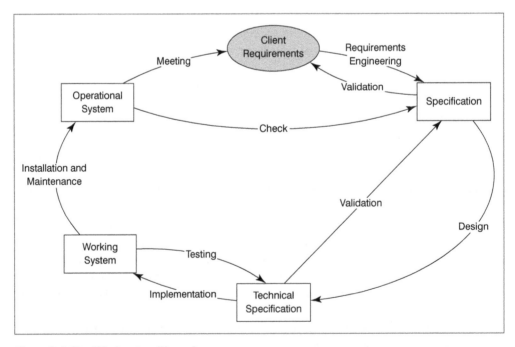

Figure 2.4 Simplified system life cycle

the data from the old system is successfully taken on by the new system. There is often a hand-holding period before the new system is formally handed over to the client. *Installation* is sometimes listed as a separate stage in the life cycle. After installation, the system development team will be involved only in maintaining and modifying the system.

Maintenance

The *maintenance* stage starts as soon as the system is formally handed over to the client. The term 'maintenance' is often used as a euphemism for finding and correcting errors that were not detected before the system was handed over. True maintenance is *modification* of the system to meet evolving client requirements. In either case, the system developer must start again at the beginning of the cycle by ascertaining the client requirements. Figure 2.4 models a simplified system life cycle as a completed circle, starting and ending with client requirements. Note also the backward-pointing arrows at each stage, which indicate the iterative nature of system development. At each stage the system development team must check back to the requirements specified in the previous stage and re-do the work if necessary.

2.2 Development methods

A software development *method* is usually based on a life cycle model of system development and has a number of development phases with a set of steps and rules for each phase. Whereas a life cycle coarsely partitions the development of a system into stages, a development method takes a life cycle and further divides each of the stages into a number of steps. A development method will prescribe in great detail what tasks are involved in each step, the nature of each task, the order in which the tasks need to be done, what documents are produced at each stage and what documents are required as *input* to each stage. In fact, it provides a detailed plan for producing a system.

Until relatively recently, development methods tended to be referred to as 'methodologies', although the word *methodology* actually means 'the study of methods'. You may still find mention of methodologies in some texts, but in this book we talk about structured approaches to developing systems as 'development methods'.

Why do we need a development method?

Comparisons are often made between the stages involved in designing and building software and other artifacts such as houses or bridges. Relatively speaking, methods for developing software have not been around for very long. However, they embody knowledge and wisdom gained by system developers through trial and error over a number of decades. For example, it is now generally accepted that one of the first and most important tasks in system development is to find out exactly what the new system is required to do. This may involve a detailed and time-consuming study of the existing system. Experience has shown that failure to do this successfully is what causes the types of error that are hardest to find and most expensive to correct. Most development methods today incorporate this experience in the requirements engineering stage of development by prescribing a set of tasks designed to ensure that client requirements are successfully elicited, recorded and validated.

Building a system involves constructing many different models of it (see Section 4.1 on modelling). Each model is only a partial description of the system. To understand the whole system, we have to understand how the models relate to and complement each other. A development method provides a framework, an agreed structure, in which these models can be related to each other.

A development method provides inexperienced system developers with a recipe to follow. Some methods are specifically designed for this purpose. Each step along the development process is prescribed, as are the ingredients, i.e. the required documents for this step and the nature of the **output** from this step.

Building a system, especially a large and complex system, is a long and complicated process during which a great number of tasks have to be done. Many of the tasks are interdependent; often, large teams of developers will be involved and their work will need to be co-ordinated. Using a method helps with the management of the whole project by breaking down the development process into small tasks, specifying the order in which they should be done and the interdependencies of the tasks. This helps with planning, scheduling and monitoring the progress of the system.

Different development methods

There are many different development methods, incorporating many different views on how to go about building a system. Different methods are suitable for developing different types of system and place emphasis on different aspects of system development. Some concentrate on the flow of data through the system, modelled using data flow diagrams, while others consider the structuring and interrelationships of the stored data in the system, modelled using entity-relationship diagrams, to be of primary importance. Yet another group is based on the prototyping approach to developing systems (see Chapter 11).

Software development methods are constantly changing to accommodate new technological advances and new ideas in system development. For example, **CASE** tools, **object-oriented system development**, and changes in client attitudes to system development, have all led to radically different approaches to software development.

There are still many practitioners who firmly believe that the way to produce perfect software consistently is to find the right development method and make sure the developers stick to it. However, there is a growing body of system developers who believe that to follow blindly a rigid method is not appropriate, given the diversity of systems tackled today. They advocate instead a tool-box approach. A craftsman asked to design and build a table will select certain tools for the job and will go about the task in a certain way. If asked to make a picture frame, the craftsman will select some of the same tools and some different ones. The process of constructing the frame is also different in many ways from that of building the table. The system developer, it is believed, should be armed with a tool-box of techniques that might be helpful for developing systems. For each new application, appropriate techniques should be used in an appropriate way. Sometimes, an entity-relationship model may be an appropriate starting point, while at other times a data flow diagram may be useful; sometimes both may be used. However, the choice of techniques should be determined by the nature of the problem, not predetermined by a development method.

Summary

The system life cycle approach was developed in response to the software problems of the 1960s and 1970s. It partitioned the development of a system into predetermined stages, each of which had to be completed and agreed with the client before progressing to the next stage. The particular contribution this brought to improving the quality of the delivered system was in concentrating more attention on capturing client requirements. It also allowed much more effective project management. Progressing from the coarse partitioning of the life cycle, development methods refine each of the stages into a prescribed series of activities with precisely defined inputs and outputs. Using a method for system development has tremendous advantages in terms of improved communication between clients, system developers and project managers. However, using a single development method to tackle very different types of system can prove to be an inflexible and inappropriate approach to system development.

Exercises and topics for discussion
What can you remember?

You will find all the answers in the chapter

a) What was the software crisis?

b) What are the stages of a typical system life cycle?

c) How does a system life cycle help project managers?

d) What is a deliverable?

e) What is included in a typical problem definition?

f) What aspects of a development project are investigated by the feasibility study?

g) What is the main difference between the current physical, current logical and required logical models?

h) What is the main deliverable from the design stage and what does it include?

i) What type of activity is generally referred to as 'maintenance'?

j) How is a development method different from a life cycle?

2.1 Why does the development of large and complex systems need a very structured approach?

2.2 List the documents you would expect to examine during the requirements engineering stage of the Just a Line system. You will find it helpful to read the interview with Sue and Harry in Chapter 3.

2.3 List some of the advantages and disadvantages of using a development method.

2.4 Draw up a table that summarizes the stages of the life cycle and the deliverables at each stage.

2.5 Some large retailers have shown an interest in the Just a Line cards. Amend the problem definition in Figure 2.2 to show the extra problems this may bring and how the system may solve them.

Further reading

Kendall, E.K. and Kendall, J.E. (2004) *Systems Analysis and Design*, 6th edn, Prentice-Hall, a division of Pearson Education, NJ.

Pfleeger, S.L. (2001) *Software Engineering, Theory and Practice*, 2nd edn, Prentice-Hall, Upper Saddle River, NJ.

Skidmore, S. and Eva, M. (2004) *Introducing Systems Development*, Palgrave Macmillan.

Sommerville, I. (2004) *Software Engineering*, 7th edn, Addison-Wesley, Wokingham.

CHAPTER 3
THE SYSTEM
REQUIREMENTS

In this chapter we look at:

- different types of requirements

- the stages of requirements elicitation, specification and validation

- interviews

- questionnaires

- structured meetings

- how to document requirements

- Fagan inspections

After studying this chapter and working through the exercises you will be able to:

- explain what is meant by the term 'requirements engineering'

- give details of the stages of requirements elicitation, specification and validation

- describe some of the different techniques used at each stage

- plan an initial interview with a client

- design a simple questionnaire

- document requirements

Introduction

The first task for a system developer in any development project is to find out as much as possible about the client organization and its problems. There are many ways of doing this, such as observing the company at work and studying relevant documents, but often the most effective way is to talk directly to the clients. In this chapter we include an interview with Harry and Sue of Just a Line that gives further details about their problems and what they hope a software system can do for them. We define and describe the **elicitation**, specification and **validation** stages of requirements engineering and explain how they would be used during the investigation of the Just a Line system.

3.1 Background

Anecdotal evidence suggests that errors in requirements may account for approximately 50 per cent of the total cost of debugging a software system, yet it is only relatively recently that serious research has been carried out on the subject of system requirements. Many of the traditional system development methods, which are generally underpinned by a standard life cycle model, merely pay lip-service to the problems of identifying, describing and validating the client's requirements for the system.

Chapter 2 describes the initial problem definition of a system. In the past this was agreed and signed off by the client, who would then frequently have little more to do with the development process until delivery and installation of the final system. Needless to say, this method of development often led to unsatisfactory systems and unhappy clients. Today, **requirements engineering** is recognized as a crucial stage in the development of software. Each year more and more requirements methods become available, often as expensive commercial products, involving computer-based tools, training programmes and extensive documentation. On the academic side, requirements engineering has become a major research area, supported by journals, specialist groups and international conferences.

3.2 What are requirements?
The problem of definition

Requirements engineering is generally regarded as having three distinct, but overlapping, stages: **requirements elicitation**, **requirements specification** and **requirements validation**, but one of the main problems in this area is that there is no consensus of opinion as to what is meant by these terms. Some developers and writers talk about requirements, others about constraints, while some make a distinction between system requirements and software requirements. For the purposes of this chapter we shall assume that *system requirements* refers to the client's needs and wishes, whereas *software requirements* covers constraints put on the system development, such as hardware, software and design methods. When capturing system requirements, the developer works with clients and users (in the case of Just a Line, this means Harry and Sue), identifying their needs, wishes and available resources, and must produce documents that can be understood by them as well as by computer professionals. When capturing software requirements, the developer works from the system requirements documents, producing software requirements that must be understood by software designers and ***programmers***. We shall also assume that the expression ***requirements engineering*** covers the three phases of elicitation (identifying requirements), specification (describing them in an appropriate language or notation) and validation (checking with the client that the description accurately records his or her needs and wishes).

Evolving requirements

Modern approaches to requirements engineering differ significantly from traditional development methods in that they do not assume that a requirements specification document is agreed by the client and then remains cast in stone for the duration of the software development project. System developers today are fully aware that requirements are dynamic and evolve constantly during development of a software system. Organizations themselves are constantly changing; although the basic business, such as selling cards for Just a Line, will remain the same, the company's scope and objectives will change. For Just a Line, for example, the ***Internet*** and the ***World Wide Web*** will have a huge impact on the way that business. Moreover, development of the software system itself has an effect on the organization; in-depth discussions about the company and its problems often lead to fresh ideas and to new ways of working that will, in turn, have an effect on the original system requirements. In the Just a Line interview that you will find later on in this chapter, Sue admits to 'an overall feeling of being disorganized'. It is likely that close examination of current working procedures during development of the new software system will make both Harry and Sue aware of where these can be improved and that this will have an effect on their original list of requirements.

Different types of requirements

In the past, requirements engineering meant defining ***functional requirements***: what the system was to do, what its inputs and outputs were and how these were linked. Correct functional requirements are still considered essential for successful software development today, but in recent years developers have also come to realize the importance of ***non-functional requirements***. These can be defined as the attributes of the system as it performs its job, and can be divided into non-functional requirements of the system and non-functional requirements arising from external sources. Non-functional requirements of the system include the following.

- **Usability.** Does the system attract or put off its intended users? Is the right level of help provided? Are options clearly displayed and easy to follow? Does the system fit in with the user's preferred way of working?
- **Performance.** Does the system respond quickly enough for the user's needs? Can it cope efficiently with the required volume of transactions? What will happen in the case where the volume of data exceeds the specified capacity of the system?
- **Reliability.** Can the client have confidence that the system will behave consistently as expected?
- **Security.** How easy is it for non-authorized users to access the system and read or modify confidential data?

Non-functional requirements that arise from external sources include methods of operation, such as the client's existing procedures, physical constraints, such as the layout of the accommodation available, and international quality control standards, such as ***ISO 9001***.

3.3 Stages in engineering the system requirements
Requirements elicitation

Requirements engineering begins with the task of finding out as much as possible about the client's organization, their current problems and what they would like the new system to do for them. This seems deceptively simple, since it involves sifting through large amounts of information and deciding what exactly is relevant. It is actually extremely difficult for the system developer to be sure that he or she has a complete and accurate understanding of what the clients want. Good communication skills, both oral and written, are essential for requirements elicitation, since nearly all methods of fact-finding depend on communication with clients and users. Requirements elicitation covers several different types of activity, such as observation of the users at work, and a study of relevant documents and user questionnaires, but the most effective way of getting information is simply to talk to the people involved in the system.

Interviews

A useful interview is one that has been prepared thoroughly. Both the developer and the interviewee should be clear about the purpose of the interview and what they each want to get out of it. A plan for the interview should be prepared in advance by the developer, identifying the purpose of the interview, listing any documents that are to be made available, and setting out a draft agenda. A plan, prepared by Mark Barnes of D&B Software, for the first interview with Sue and Harry Preston of Just a Line is shown in Figure 3.1.

The developer should have established relevant details about the interviewee, such as his or her background, position in the organization, length of time with the company and special skills, such as level of computer expertise. It is part of the developer's job to put the

D&B Software – Interview Plan			
System: Just a Line		**Project reference:** JaL/MB/02	
Participants: Sue Preston (Just a Line) Harry Preston (Just a Line) Mark Barnes (Developer)			
Date: 10/4/05	**Time:** 14:30	**Duration:** 45 minutes	**Place:** Prestons' house
Purpose of interview: Preliminary meeting to identify problems and requirements regarding a system for the *Just a Line* card retail company.			
Agenda: • current problems, particularly customer ordering, and any other concerns • current procedures • initial ideas • follow-up actions			
Documents to be brought to interview: • current catalogue • any documents relating to current procedures			

Figure 3.1 Plan for the first interview with Sue and Harry from Just a Line

interviewee at ease, particularly if the interview takes place away from the interviewee's place of work. It is worth spending some time chatting to the interviewee in general terms and very important to listen carefully to what he or she has to say, even if it does not appear to be directly relevant. The ability to listen attentively and identify important and relevant information is one of the essential skills for a system developer during requirements capture. Although direct questions are needed to control the interview, a lot of information can also be discovered by smiling, nodding encouragingly and making the interviewee feel that what he or she is saying is important. The developer should direct the interview, but must not dominate it.

On the next few pages you will find the first interview between the system developer and Harry and Sue of Just a Line. In any conversation of this type we can find several different kinds of information. Some of these are listed below.

- Information that is already structured in lists or forms.
- Information about company procedures; how certain tasks are carried out at present.
- Measurements such as the number of customers or the average size of an order.
- Problems that the client has identified in the current system. Definite requirements for the new system.
- Information that is not stated directly, but where there are definite 'vibes'. An example of this might be where the clients complain that they are always rushed when the supplier's order comes in, whereas what is actually happening is that the supplier always delivers late.

As you read the interview with Harry and Sue, try to identify examples of the different kinds of information that tell the system developer about Just a Line.

Interview with Sue and Harry from Just a Line

HARRY Nice to see you. Traffic a bit heavy, was it?

MARK BARNES (SYSTEM DEVELOPER) Yes, dreadful. I'm sorry I'm a bit late, I got badly held up leaving the motorway.

SUE Don't worry, it was good of you to phone and let us know. Now, a cup of tea perhaps. I'll put the kettle on and we can tell you a bit about the company.

MB Yes, I've heard a lot of good things about your cards, and I was wondering if you could show me some samples later on. I'm very interested to hear how you got the idea for the business.

SUE Of course we'd love to show you the cards and we're always keen to get comments on possible new lines.

MB Well, just to put you in the picture about this meeting – what I'd like to do is get a good idea of the company and what you'd like from a computer system. I'd like to cover what you do, how you work, what you see as your current problems and how you think a software system could help. Is that OK with you?

HARRY Fine, Sue will be quite happy to talk about Just a Line all night.

MB Right, perhaps you could start by describing the main jobs you have to do to run Just a Line. What about the first contact with the customers, for example?

SUE Well, we meet them when we take the orders, or speak to them at least, if it's a phone order – it usually is. That's a real pain. Most of the customers are so chatty, they ask all about what we do and what we stock, and we have to explain all about the different sorts of cards and designs and all that. We try to persuade them to come into the shop to see for themselves, but they don't always want to. Then we get on to the personalized bit and we have to talk them through all the various typefaces and colours and what have you. People are always so friendly and interested and they usually end up ordering quite a few cards, but it's terribly awkward to deal with it all over the phone.

MB What happens when they ask you detailed questions? Do you have all the information in your heads, or do you have to go and look things up all the time? And perhaps you can tell me how you keep information about card designs, prices, etc.

SUE We've got a list of the card designs we offer on a regular basis – there's a copy stuck on the wall by the phone. We can usually remember what the design or picture is like by the name, so we just describe it to them as best we can, and quite often people have seen samples on our flyers in places like the library anyway.

MB Do you get many comments from customers about your ordering procedures – either good or bad?

SUE Well, our local customers do like the personal touch, but I'm sure a lot of people find it very frustrating that our processes aren't more streamlined.

MB I think it might be a good idea to ask customers to fill in a simple questionnaire about ordering cards from Just a Line. What do you think?

HARRY Yes, that's a good idea. It would give us more information about how the customers feel about the way we handle that side of things.

MB Good, we can get that organized. Now, what about the prices? Is that a separate list?

SUE Yes, we keep that by the phone as well, so that it doesn't get lost. The price list gets updated more often than the design list – in fact here's an old one; you can keep it.

MB I see. I suppose this is how you hold information about suppliers, too. I see it's got their name and address on it, as well.

SUE Yes. Well, at the moment there's only one supplier. We're thinking about using a different one who does recycled paper. We're very keen on that. But we may be able to persuade our present supplier to sell it as well, which would be a lot simpler than dealing with two suppliers. Anyway, eventually the customers tell us what they want to order and we take it down, in triplicate, on our order forms using carbon paper.

MB Three copies?

SUE The top two copies go out with the order. The customer keeps one as a delivery note, the other comes back with the customer's payment. We keep the third one until the signed copy comes back with the payment. Everything's cash on delivery at the moment, although we realize we're going to have to start thinking about credit accounts and that sort of thing.

MB What about pricing? Do you work out the cost when you take the order?

SUE Sometimes, because customers want to know how much it's going to cost. But quite a lot of people just leave it to me to work out the cost later. That's better really.

MB Then what happens?

SUE To the orders? Well, they go in the order file – our whole filing system is just one big filing cabinet with three drawers and orders go in the top drawer. They go in the drawer in the order we receive them, unless there's a very urgent one. We generally tell the customers to allow a maximum of seven days before they get the cards.

MB Then, when the cards are ready you deliver them?

SUE Yes, if it's local – or else we use the post.

MB Well, I've got a reasonably good picture of how you handle orders. Just one other thing – have you thought about selling cards via the Internet, you know with a website to display the designs, and online ordering?

SUE Yes absolutely, I think it's a brilliant idea. I'm not sure that Harry's so keen though.

HARRY Well, I'd just rather get a straightforward system going here first. I know the Internet seems to be a perfect way of selling the cards, but it would mean a huge amount of change in the way we do everything and I'm not sure that we're ready for that.

SUE But Harry, you know how important it is to keep ahead of the game. I think our designs are great, but they won't sell themselves without some sharp marketing.

MB Well, let's leave that for the moment; we can always come back to it. What we can do is design this system so that we can develop Internet facilities later on if you decide that's the way you want to go. What about dealing with stock? How do you organize that?

SUE We usually reorder about once a month. Our current supplier delivers free if we can order more than £500 worth of stuff. At first we couldn't always afford to make that saving, but things are going reasonably well now, so we usually manage to. It's complicated. We can't just order one particular line if we've got a run on it, but we usually manage to organize a large enough order each month. Fortunately, we can get what we order delivered within three days. When it's obvious we're getting low on a few lines, I check to see if we can order enough to get the free delivery. We swore we'd never run into debt with the supplier, and so far we've managed to keep to that.

MB How do you actually decide what to order?

SUE It's a bit hit and miss sometimes. I look in the store where we keep all the cards and make a list of what's getting low, particularly with popular lines. Then I have a look through the orders from customers and make an estimate of what we need. The idea is to order so that everything runs down at about the same rate. I've pretty well got the hang of it by now.

MB OK ... so you have to keep checking manually that you have enough cards of each line in stock to cope with current orders and what you think customers will order in the near future? As you say, it sounds as if it can be a bit hit and miss, but it gives me a good idea of how a new system will be able to help. We can make sure that the system keeps track of all sales, and alerts you when an item is running low. By the way, what happens if you do run out of something and can't meet a customer's order within the seven days?

SUE We just tell the customers we haven't got what they want at the moment. They usually don't mind as long as we tell them when they order. If we find out later, then we phone them. They're mostly very nice about it. I get Harry to tell them, he's much better at that sort of thing than I am and he can often persuade them to order something else instead. We always try to remember to say everything is 'subject to availability' when they place the order, but it's easy to forget.

MB So, you send out orders to your suppliers a bit irregularly, but roughly once a month. What happens when the stuff arrives? How do you pay?

SUE They send an invoice with the goods. We check everything's there and sign for it. Then they invoice us. Then we pay them.

MB What is it you sign, when they deliver?

SUE It's the bottom copy of their delivery note.

MB You have to record customer payments, too, of course?

SUE Yes. When we get a payment we take our copy of the order out of the top drawer of the filing cabinet and put it in the second drawer. We call that one 'past orders'. We make a note of the payment in our cash book. Harry usually banks the money weekly, or daily if a lot has come in.

MB Cash book? What else goes in there?

SUE Everything, I'm afraid! We record all the money going in and out. So, as well as customer payments, we record payments we make to our supplier, and all our running expenses, like petrol, postage, phone calls, and so on.

MB Well, I think I've got some idea about how you run the company, now. My main impression is that we definitely need to work on the customer ordering side. I'm sure we could make that a lot more efficient. What I'd like to do is go away and sort it all out a bit in my mind, then come back to you and check I've got things right. In the meantime, are there any other problems you feel we haven't covered?

SUE Not really, it's just an overall feeling of being disorganized. That worries me; especially now that some of the big stores seem to be interested in the cards. I do think we've got a good product to sell and I don't want to mess it all up by seeming to be unprofessional.

MB Well, that's just the sort of problem we deal with. I'll be in touch again soon and I'm sure we'll be able to get you organized. Now could I have a look at the cards? I'm really keen to see them after all you've told me.

HARRY Great, that's terrific. If you'd like to order some I could even do you a discount.

Questionnaires

Apart from interviews with clients and users, the most useful form of requirements elicitation is often a questionnaire. This is particularly effective when a small, well-defined amount of information is needed from a large number of people, especially if they are widely scattered. It could be used, for example, in the Just a Line case study to find out what the company's customers think about the Just a Line method of ordering cards.

As with interviews, it is essential to prepare questionnaires thoroughly, including testing on a small sample of people to ensure that the questionnaire is easy to understand, simple to fill in

---

Just a Line Ordering Service – Customer Survey

We are intending to move to a new computerized ordering system for our cards in the near future. It would be a great help to us if you could spare a few minutes to give us your opinions of our current ordering system and any suggestions you have for improving it. Please answer the questions below and return the form to us in the enclosed pre-paid envelope.

1. How many times have you bought cards from Just A Line?

    Once only   [ ]

    2–5 times   [ ]

    6 times or more [ ]

2. Have you experienced any problems with the Just a Line ordering system?

    Yes   [ ] please explain briefly below

    No   [ ]

_____

_____

3. For each of the statements (a)–(g) shown below, circle the number that is closest to your own view, where 1 means that you agree strongly with the statement and 5 means that you disagree strongly.

    (a) The current system works well.   1 2 3 4 5

    (b) The staff are always friendly and helpful.   1 2 3 4 5

    (c) It is easy to choose and order cards.   1 2 3 4 5

    (d) I would like to see more information about my order.   1 2 3 4 5

    (e) I feel that a computerized ordering system would be more efficient.   1 2 3 4 5

    (f) I would like to view card designs on the web.   1 2 3 4 5

    (g) I would like to order card designs on the web.   1 2 3 4 5

4. Please note below any other comments you have on the current system.

_____

_____

_____

5. Please note below any suggestions for the new system.

_____

_____

_____

Your name: _____

Your address: _____

Thank you for completing this questionnaire.

Figure 3.2 Just a Line ordering service – customer survey

and that it will produce useful results. It is the responsibility of the system developer to make sure that people who fill in the questionnaire are aware of its purpose and how their answers will be used. A variety of question types may be used, including multiple choice, short answer and extended answer questions, but the main priority must be to ensure that all questions are as clear and straightforward as possible. If a question does not contain enough information, the person filling in the questionnaire will not understand what is required, but if it contains too much information, nobody will bother to read it.

Figure 3.2 shows a questionnaire on Just a Line's ordering methods. The purpose of the questionnaire is to help the system developer find out what Just a Line's customers think about the way ordering of cards is handled at present and to elicit ideas from them on how the process might be improved.

Structured meetings

Just a Line's is a relatively small, simple system and requirements elicitation could be carried out using just interviews and questionnaires. However, in the case of large, complex systems, successful requirements elicitation often involves structured meetings with **stakeholders**, who may include not only clients and users, but also user managers, marketing staff, project managers and software trainers. Such meetings often take up the time of many people and so must be carefully planned with a well-defined purpose and clear agenda. Effective management during the meeting is also important in order to avoid unnecessary conflict or deviation from the main purpose.

One particular type of structured meeting, which originated in Scandinavia, is known as the future workshop. Future workshops come from a development approach called Participatory Design, which views users of the system as experts in the problem domain and aims to include them as active collaborators in the development process. A future workshop is generally organized and run by two facilitators who spend some time beforehand familiarizing themselves with the client organization, its processes, and any existing hardware and software.

The workshop itself is divided into three separate stages: critique, fantasy and implementation. During the critique stage, participants in the workshop focus on current problems in the organization; brainstorming rules apply, so speaking time is limited, statements do not have to be justified and personal criticisms are not allowed. The aim of the fantasy stage is to imagine the perfect future system, without considering any constraints. This, again, is carried out using brainstorming to generate statements about the future system; at the end of this part of the workshop, participants vote to rank the statements, in order to identify those that have most support. In the final part of the workshop, the implementation stage, participants discuss ways in which the fantasy view may be realized, taking recognized constraints into account. After the workshop, the facilitators develop part of the proposed implementation as a prototype in order to validate and refine the system requirements.

Requirements specification

Whereas requirements elicitation involves an expansion of the developer's knowledge about the problem domain and the client's wishes, requirements specification involves sifting through the information to filter out the important and relevant issues and record them in an appropriate form. This may be narrative English, diagrams or a mixture of the two. The techniques described in Chapters 4, 5, 6 and 7 are commonly used to record requirements.

Alternatively, in certain *safety-critical* or *security-critical systems*, it may be appropriate to describe some or all of the requirements using a formal, mathematical language such as the *Z specification language*. It is beyond the scope of this book to discuss such languages, but you can find references in the bibliography to books that describe them in detail. One of the most effective ways of recording requirements for a software system is *prototyping*, where the developer builds an unpolished version of all or part of the system. Clients and users can then get a feel for what the new system will be able to do and what it will look like. Prototyping allows clients to see how their requirements translate into a computer system; it is particularly useful when requirements are uncertain. You will find a discussion of prototyping in Chapter 11.

Whatever language or method is chosen for the specification of requirements, certain information must be provided and the requirements specification itself must have certain qualities. For each separate requirement the following information should be included (see also Figure 3.3):

- a number or code that uniquely identifies the requirement
- the source of the requirement
- the date this version of the requirement was suggested
- a brief, natural-language description of the requirement
- the priority of the requirement (essential (E), desirable (D) or optional (O))
- a list of related requirements
- alternatives to the requirement, if any
- related documents, diagrams or tables.

If the new requirement involves a change to a previous one, then this must be fully documented, together with the reasons for the change and the effects it will have on other system requirements.

The specification is a cornerstone of a system development project, since it encapsulates the shared understanding and intentions of all the stakeholders. The specification may be used as a vehicle for communication between developers, users and other stakeholders; it may also form the basis of a legal contract between developer and client, and it is the document that guides the programmers in their implementation of the system. Much has been written recently about the desirable qualities of a requirements specification. One of the most useful sources is the IEEE Recommended Practice for Software Requirements Specifications (IEEE, 2000). The standard describes qualities that are essential for a good requirements specification document, including correctness, *consistency* and understandability. Many of these qualities are simply common sense, but the Standard is useful as a checklist.

No.	Source	Date	Description	Priority	Related reqs	Alternative reqs	Related docs	Change details
4.9	Meeting with Sue and Harry Preston	10/4/05	System must keep track of sales and alert when an item is running low.	Essential	2.9 4.6		Sue's lists of current stock and estimates of what needs to be ordered.	

Figure 3.3 Specification of a requirement from the Just a Line system in tabular form

Requirements validation

Although it is a time-consuming process to check that a requirements specification has all the qualities mentioned above, it is technically feasible for the system developer to confirm that the requirements specification document is actually of the desired quality. What is much more difficult to ascertain is whether the requirements expressed in the specification are really what the client wants and needs. The situation is further complicated by the fact that the client may not know what he or she wants, or that what is wanted may be completely different from what is needed. The process of checking that the requirements as specified are a true representation of the client's needs and wishes is known as *validation*.

Validation of requirements is essential from the earliest stages of requirements engineering. During interviews with clients and users there should be constant feedback to ensure that the developer has fully understood what is being said. This is also useful in that it helps the interviewee to feel that what he or she is saying is helpful and relevant. The extract below from the Just a Line interview illustrates these points.

> **SUE** It's a bit hit and miss sometimes. I look in the store where we keep all the cards and make a list of what's getting low, particularly with popular lines. Then I have a look through the orders from customers and make an estimate of what we need. The idea is to order so that everything runs down at about the same rate. I've pretty well got the hang of it by now.
>
> **MB** OK ... so you have to keep checking manually that you have enough cards of each line in stock to cope with current orders and what you think customers will order in the near future? As you say, it sounds as if it can be a bit hit and miss, but it gives me a good idea of how a new system will be able to help. We can make sure that the system keeps track of all sales, and alerts you when an item is running low.

Initial validation is also carried out by taking notes during the interview and later producing a written summary for the interviewee. The developer should always ask permission to take notes and be prepared to show the interviewee what is in them. The written summary should be produced shortly after the interview, so that the interviewee remembers what was said, but has had time to think about it and can check that the developer has understood the important points. A summary of the initial interview with Sue and Harry from Just a Line can be seen in Figure 3.4.

Different methods of requirements elicitation can be used in conjunction to validate requirements. This may be carried out by comparing answers on a particular topic from a questionnaire with comments on the same topic that have been obtained during interviews. A client's account of certain business procedures may also be checked by observation of how the procedures are carried out in practice. Other techniques may also be introduced to aid the validation process. The technique of prototyping, discussed in Chapter 11, is useful at all stages of requirements capture. When used for validation, a prototype allows the client and users to get some feel for how their ideas would work once implemented in a computer system. The structured modelling techniques described in Chapters 4, 5, 6 and 7 are designed to be user-friendly and to facilitate validation. Since the client's original requirements are generally expressed in natural language, one of the most effective methods of validating the requirements specification is simply to talk through it with the client and users.

D&B Software – Interview Summary			
System: Just a Line		**Project reference:** JaL/MB/05	
Participants: Sue Preston (Just a Line) Harry Preston (Just a Line) Mark Barnes (Developer)			
Date: 10/4/05	**Time:** 14:30	**Duration:** 45 minutes	**Place:** Prestons' house
Purpose of interview: Preliminary meeting to identify problems and requirements regarding a system for the Just a Line card retail company.			

No.	Item	Action
1.	Phone orders are time consuming and complicated, both for Just a Line and probably customers.	Produce simple questionnaire on current ordering procedures and ask customers to fill it in with next order.
2.	Designs often have to be described over the phone.	Consider setting up a website to display designs.
3.	Prices have to be updated manually on a central list.	
4.	Possible move to more than one supplier.	New system must cater for multiple suppliers.
5.	Currently no facilities for customer credit.	
6.	Total cost of order cannot always be given over the phone.	
7.	Sometimes orders cannot be filled because of stock problems.	New system must keep track of sales, and alert when stock of an item is running low.
8.	'Overall feeling of being disorganised'.	New system must streamline procedures to make company more efficient. Arrange follow-up meeting with Sue and Harry (in about 10 days' time).

Figure 3.4 Summary of the initial interview with Sue and Harry Preston of Just a Line

One very popular way of validating requirements is to carry out a **_Fagan inspection_**. Fagan inspections aim to uncover defects in the output from any stage in the software development process; they are a systematic and structured way of checking documentation, such as the requirements specification. Each inspection usually concentrates on a small component of the documentation and lasts up to two hours. Inspections are carried out by small teams of people consisting of a moderator or chair, a note-taker, the person who produced the component and one or more people to inspect it. A typical Fagan inspection has six separate stages:

1. planning, when the team is selected
2. overview, when the component is presented to the team
3. preparation, when each team member studies the component individually
4. meeting, when the team members review the component together
5. rework, to remedy the defects that have been identified and agreed by the team
6. re-inspection, which may be carried out by the chair rather than by the whole team.

Although it is never possible to prove conclusively that a requirements specification describes exactly what the client wants and needs, it is essential that both the system developer and the client are happy that the requirements as stated have been thoroughly validated. The validation process can be seen as the application of *quality assurance* concerning the requirements specification, leading to a well-founded belief on all sides that the specification is an accurate description of the system requirements.

Summary

Requirements engineering is one of the most important stages in the system development process, since getting the requirements wrong can (and often does) cause serious problems. Requirements engineering is generally considered to have three distinct, but overlapping stages. These are requirements elicitation (where the developer tries to find out as much as possible about the client's problems and what they want from the new system), requirements specification (where the requirements are recorded as fully and as accurately as possible), and requirements validation (where the developer checks that the client's requirements have been understood and described correctly).

Exercises and topics for discussion
What can you remember?

You will find all the answers in the chapter

a) What are the three stages of requirements engineering?

b) What is meant by 'evolving requirements'?

c) What is the difference between functional and non-functional requirements?

d) Give two examples of non-functional requirements.

e) What happens during requirements elicitation?

f) What is one of the most important skills in requirements elicitation?

g) Name three of the techniques used in requirements elicitation.

h) What happens during requirements specification?

i) Name two ways of describing requirements.

j) What happens during requirements validation?

k) What is an interview summary used for?

l) What is a Fagan inspection?

3.1 Find an example from the Just a Line interview of each of the different types of information listed in Section 3.3.

3.2 What other questions could the system developer usefully have asked Harry and Sue?

3.3 You have been asked by the town council to develop a system to regulate traffic in a local car park. Design a questionnaire that will help you to find out how the car park is used at present and how people feel that it could be improved.

3.4 Organize a future workshop with a small group of people to identify initial requirements for one or more of the following systems:

- a cash machine to be situated on a university campus
- a system to handle orders from a charity's Christmas catalogue
- a public information system for your local area
- a system to control traffic lights at the junction near a village school.

The workshop should cover the three stages of critique, fantasy and implementation, as described in this chapter in the section on structured meetings.

Further reading

IEEE (2000) *0830–98 Software Requirements Specifications*, Institute of Electrical and Electronics Engineers, Inc., New York.

Kotonya, G. and Sommerville, I. (1998) *Requirements Engineering: Process and Techniques*, John Wiley & Sons, Chichester.

Pfleeger, S.L. (2001) *Software Engineering, Theory and Practice*, 2nd edn, Prentice-Hall, Upper Saddle River, NJ.

Sommerville, I. and Sawyer, P. (1997) *Requirements Engineering: A Good Practice Guide*, John Wiley & Sons, Chichester.

CHAPTER 4
PROCESS
MODELLING

In this chapter we look at:

- modelling in software development

- the modelling techniques used in structured software development

- data flow diagrams

- the aspects of a system modelled by data flow diagrams

- how data flow diagrams change as the system evolves

- process definitions

- structured English

- decision trees

- decision tables

After studying this chapter and working through the exercises you will be able to:

- explain why software developers use models

- explain the aspects of a system that are described by each of the various modelling techniques

- interpret, draw and check a data flow diagram

- explain the difference between current and required data flow diagrams, and between physical and logical data flow diagrams

- decide when to use a process definition

- describe a process using structured English, a decision tree or a decision table

Introduction

In Chapters 4, 5, 6 and 7 we discuss a range of modelling techniques used by most structured system development methods in the requirements engineering stage of system development. The techniques introduced are **process modelling**, **data modelling**, **event modelling** and **data dictionary**. Process modelling includes **data flow diagrams** and **process definitions** (see Figure 4.1). All information systems *do* something; typically this involves the input of data to the system, which is then processed to produce the system outputs. For example, orders input to a mail-order system are processed, i.e. checked, priced and made up. This results in the system outputs: an order delivery to a customer with a delivery note followed by an invoice. Data flow diagrams are useful for modelling the system processes and the data input, processed and output. Data modelling includes E-R diagrams and normalization. Data modelling concentrates on the stored data requirements of the system – the data that the system needs to store in order to function, and the most efficient way of storing it. Event modelling includes entity life histories and state diagrams. Both of these techniques look at the system entities (such as customers, orders, products) in terms of events that can affect them and how they respond. The data dictionary is used to document all of the other models – for example, to describe the content and structure of **data information flows**, **data stores** and **entities**. Data flow diagrams and process definitions are discussed in this chapter; data dictionary is discussed in Chapter 5; E-R diagrams

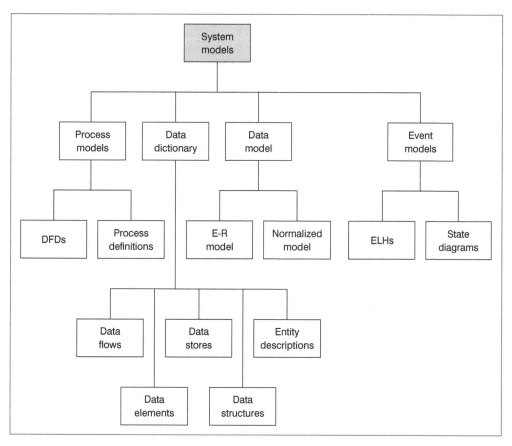

Figure 4.1 Structure of modelling techniques

and normalization are discussed in Chapter 6; and entity life histories and state diagrams are discussed in Chapter 7. The Just a Line system is modelled using each technique.

Each technique provides its own view of the system and concentrates on this view while ignoring other aspects. These perspectives are explained and illustrated. When developing a system it is a matter of choice whether to do process modelling before data modelling or vice versa – sometimes it is useful to do them simultaneously. Entity life histories and state diagrams use information provided by both the data model and the process model and must, therefore, be done after them. As each model is developed, it should be documented in the data dictionary.

This chapter begins with a discussion of modelling in general and why modelling is used in system development. The comments on modelling apply to all of the modelling techniques covered in this book. The chapter then goes on to discuss data flow diagrams in two parts. Part 1 (Section 4.2) gives an overview of the technique, its purpose, notation and rules. This section includes the concept of levelled data flow diagrams but avoids detailed discussion of the differences between *physical* and *logical models.* Revision questions a) to q) and exercises 4.1 to 4.8 test the reader's understanding of the ideas introduced here. Part 2, in Section 4.3, discusses the difference between physical and logical data flow diagrams. Revision questions r) to t) and Exercises 4.9 to 4.14 relate to this part. Beginners may prefer to omit Section 4.3 on a first reading.

4.1 Modelling
Introduction

Modelling is used extensively in structured system development. For this reason it is important to understand what system developers mean by the term. A model of a system represents a part of the real world. The model differs from reality in that it concentrates on certain aspects while ignoring others – it is an *abstraction*. Modelling is the process of abstracting and organizing significant features of part of the real world as it is now, or as we would like it to be. A useful model will represent only those features of reality that are useful for the current purpose. Each model says something about the subject, but not everything. Different types of model illustrate different things.

The system developer is not alone in using modelling in this way. An architect, for example, is modelling when he or she draws a rough sketch of a house to show a client approximately what the house will look like. This model is different from, and serves a different purpose from, a scaled drawing of the same house, which might be drawn to show planners how it will fit in with its surroundings. Both of these models, however, are two-dimensional; the architect may also construct a three-dimensional model of the house to give the client a clearer understanding of the design, of how the house will look from all angles and how the rooms relate to each other. Plans will also have to be drawn to serve as working documents for the builders and site engineers – blueprints drawn to scale with precise measurements of widths, depths, materials and loadings specified. Each model is a partial representation of the house the architect is designing; each model says something about the building, but not everything. Each is used for a particular purpose and makes a different contribution to the development of the building.

Using models for communication

It is no accident that structured system development has introduced the extensive use of modelling, particularly graphical modelling. Earlier attempts to produce documents expressing the client requirements of the system, known variously as the requirements or functional specification, were conspicuously unsuccessful. This document was typically an enormously thick and indigestible text with little or no use of graphics to leaven the lump. Clients were supposed to 'agree' and sign it. Frequently, it was never completely read either by the clients, who therefore failed to pick up any errors, or by the system developers, who eventually had to translate it into a piece of software.

A graphic representation of a system has several advantages over a narrative document. Diagrams tend to be more readily understood by both clients and system developers. Diagrams can express ideas more concisely, which makes the size of the documentation less daunting.

System developers use models in the requirements engineering stage for various purposes. Experience has demonstrated the benefits of modelling to be as follows.

- Models impose structure on the jumble of facts and opinions gathered while discussing the system with the client. They are used both to record facts and to sort them into some kind of order.
- The system developer models the system as he or she currently understands it. Discussion of these models with the other people involved, including the clients, will identify misunderstandings. Having a model on which to centre discussions makes this process easier.
- During the construction of the models the system developer becomes aware of questions that need to be asked, details that have been left out and contradictions in his or her understanding of the system. Producing the model highlights the shortcomings; discussing the model with clients will help to resolve them.
- The same modelling techniques are used to describe the existing system and the new, required system. Clients may find some effort is required to understand the modelling techniques, but the effort will be worthwhile as the understanding gained can be applied to models at all stages of system development.
- Once modelling of the new system is complete, the system developer can test the model and check it for consistency and completeness. It can be evaluated informally and provide a fairly good idea, at an early stage in the development of a system, whether or not it can satisfy the client requirements. This sort of checking would be almost impossible to do using a purely narrative document because of its unwieldy nature.

Using models to tackle complexity

Models help the system developer to communicate with the client and help the client to understand the new system. They are also useful to the system developer when tackling the complexity of large systems. Computer scientists have commonly used two main intellectual techniques to cope with complexity:

1. **decomposition** – dividing the problem into 'brain-sized' chunks
2. **abstraction** – concentrating on the most important elements while ignoring currently irrelevant details.

The models used in structured system development make use of both these techniques. The first, *decomposition*, takes a divide-and-conquer approach. A large and complex problem will be made up of lots of smaller problems. The developer will keep splitting the problem into smaller and smaller sub-problems, until a brain-sized problem is left, i.e. one that is small enough to be handled easily. The developer's need to decompose the problem is supported, in particular, by the use of data flow diagrams (see Sections 4.2 and 4.3), which can be used to split up the problem using top-down *functional decomposition*. We will see that some of the modelling techniques used in structured system development allow us to view the problem at different levels. They offer us the ability to have both a general overview of the problem and to view, selectively, parts of the problem in detail.

The second technique, abstraction, allows a developer to concentrate on one aspect of a problem at a time. In designing a house, the architect can safely ignore, for example, consideration of the precise type of brick to be used, when deciding how many bedrooms the house will have. In the same way, the system developer can safely ignore the detail of problem-solving algorithms when deciding what the user interface will look like. The modelling techniques used in structured system development support the developer's need to concentrate on one aspect of the problem at a time in that each focuses on its own particular perspective of the system.

Different perspectives

Most structured development methods use some or all of the following modelling techniques during analysis. As with the architect's models, each is used for a particular purpose. Each model has its own view or perspective of the system – it concentrates on certain aspects and ignores others. Combining these views gives the whole picture. The list below illustrates the perspective offered by each of the techniques discussed in this chapter.

Modelling technique	Perspective
Data flow diagram	What the system does – shows how data pass through the system and affected by it
Data dictionary	Detail about the system data, e.g. contents of data flows and data stores
Process definition	Detailed textual descriptions of processes identified in the data flow diagrams
Data model	Identification of the stored data requirements of the system and the most efficient way of storing the data
Entity life history and state diagrams	Examination of how data items in the system respond to external events

Each of these modelling techniques is explained in detail in Chapters 4–7.

Different stages

During the analysis of a system the same technique may be used to model the system at different stages. Like architects' models, the models in the early stages of system design are non-technical. Later on in the project development a more precise version of these early sketches may be used as the basis for a technical design. A data flow diagram, for example, may be sketched during the feasibility study, modified to describe the existing system, then updated to describe the required system. In this form it is a non-technical diagram and can be discussed with the client. Eventually, the data flow diagram may provide the basis for the design of the system structure and program specifications. Similarly, a data model may be drawn of both the existing and required system and eventually used as the basis for database design.

4.2 Data flow diagrams: part 1
Introduction
Most structured development methods use data flow diagrams (DFDs) as one of their modelling techniques. Data flow diagrams identify the **system boundary** or **automation boundary**,[1] the external entities (see below), the data stores, and the data or information flows into and out of the system. They chart the progress of the data through the system and through the processes that affect it. Data flow diagrams are, fundamentally, **process** based. They concern themselves with the question: What does the system 'do' to the data: how does it process the data? This process-oriented view of a system may be contrasted with the object-oriented view discussed in Chapter 11. Data flow diagrams are useful because they provide a graphic overview of what the system does in a manner that is easy to understand. However, they deliberately leave many questions unanswered. Many aspects of the system are modelled using the complementary techniques of data dictionary, data modelling, process definitions, entity life histories and state diagrams.

How do data flow diagrams work?
Data flow diagrams aim to be immediately accessible to both layman and specialist alike. Figure 4.2 shows a fragment of a data flow diagram modelling part of the Just a Line system. This illustrates four of the basic elements that data flow diagrams use to model the system:

1. data flows
2. processes
3. data stores
4. external entities.

Data flows
Data flow diagrams model the flow of information or data through the system. In Figure 4.2 a customer's order is modelled as a data flow labelled **order** going from a customer into the system. The list of products and prices given to customers is modelled as a data flow from the system to the customer, labelled **productDetails**.

A data flow is modelled as an arrow with a unique label that briefly describes its content. Usually a data flow represents a group or packet of data items. When something significant happens to a data flow its label should reflect this. In Figure 4.2, for example, before orders are agreed, stock levels are checked to see that the order can be met. Once this check has been made an **order** becomes a **validOrder**; the label of the data flow changes to reflect the new status of the order.

---

[1] Some versions of data flow diagrams, for example SSADM, do not model the system boundary.

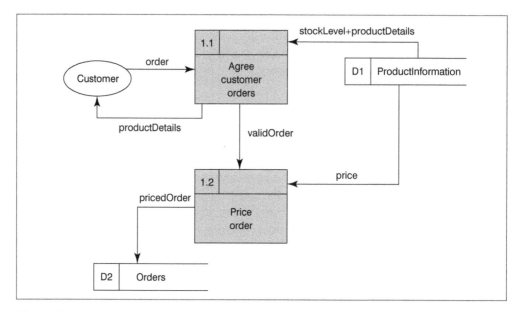

Figure 4.2 Fragment of a data flow diagram showing processes to agree and price orders. Customers are given information about the products stocked and send in orders. Stock levels are examined to check that the order can be met; if so, the order is deemed to be 'valid' and is priced and filed

Processes

A process models what happens to the data; it transforms incoming data flows into outgoing data flows, i.e. it processes the data. Typically, a process will have one or more data inputs and will produce one or more data outputs. Processes are shown on the data flow diagram as rectangles (see Figure 4.2) and are labelled with a brief description of their function (e.g. **Agree customer orders**, **Price order**). Each process has a unique reference number, in the top left-hand corner of the process box. The top right-hand corner of the box is used to document (optionally) where a process takes place or who does it. Names such as **Personnel**, **Accounts** or **Receptionist** appear here. Normally, the location or perpetrator of a process is documented only on a diagram that describes how an existing system works, i.e. a current physical data flow diagram (see Section 4.3).

Data stores

A data store represents permanent data that is used by the system and must therefore be stored if the system is to function correctly. Data stores are shown on the data flow diagram (DFD) as open-ended rectangles, each labelled to reflect its content. In Figure 4.2 there are two data stores, **ProductInformation** and **Orders**. Stock levels are recorded in the data store **ProductInformation** and examined before an order is agreed. **ProductInformation** is also used to store descriptions of the Just a Line product range and product prices. The second data store, **Orders**, is used to store priced orders until they are ready to be made up and delivered.

   In complicated situations it is sometimes useful, to avoid having crossed data flows, to repeat a data flow element such as a data store or an **external entity**. If a data store appears more than once on the same DFD it is marked by an extra vertical line on the far left of the data store symbol (see Figure 4.3). This convention does not apply to process boxes.

External entities

External entities are people, organizations or other systems – anything outside the system boundary that sends data into the system or receives data from it. They are not considered to be part of the system but are external to it. In Figure 4.2 the customer is modelled as an external entity, receiving information about products, **productDetails**, and sending in an **order**. External entities are represented on the data flow diagram as ellipses with an appropriate label. Sometimes external entities are referred to as sources and sinks; an external entity either supplies data to the system, which makes it a source, or receives output from the system, which makes it a sink, or both. External entities are drawn outside the system boundary. Repeated external entities are marked by a 'cut-off corner' symbol, see Figure 4.3.

The system boundary, the fifth basic element used in data flow diagrams, is represented by a dashed line and should be shown on all diagrams. It is not drawn on Figure 4.2 only because this is a partial data flow diagram, used to introduce the notation (the complete diagram appears later in Figure 4.6).

Data flow diagram notation

Several different notations exist for expressing data flow diagrams – there is no definitive standard. However, it is only the shapes of the symbols that vary, not the underlying logic.

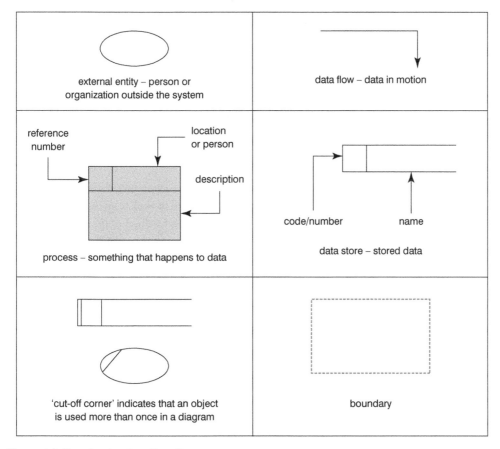

Figure 4.3 Notation for data flow diagrams

The notation used in this text follows one of the most commonly used conventions (Figure 4.3). Most varieties of notation use the elements identified above: data flows, processes, data stores, external entities and system boundary.

Levelled data flow diagrams

Data flow diagrams clearly and graphically represent the flow of data through a system. However, even a small system may have several hundred processes; the data flow diagram could cover a wall. To deal with this problem, levelled data flow diagrams are used. The idea of levelling is to allow us to view the system at different levels of detail, to offer both a general overview of the system and selective viewing in progressively greater detail. This is a familiar technique; we use maps in the same way. Maps of different scales allow us to view the whole world at one time or one country in more detail; to see the finer detail of one particular region we can use a large-scale map of the area; street plans are used to home in on the detailed layout of a town.

Context diagram (level 0)

The top-level diagram, level 0, is known as the **context diagram**. It models the whole system as a single process box whose sides represent the boundary of the system. It identifies all external entities and related input and output flows. By defining the boundary of the system, the context diagram delineates the domain of study so as to define those functions or areas of activity that are to be included and those that are to be excluded. Figure 4.4 shows the context diagram of the current Just a Line system. The diagram identifies **Customer** and **Supplier** as external entities. Customers receive information about the Just a Line products – **productDetails**; they also receive **invoices** and **deliveryNotes**. The customers also supply the Just a Line system (Sue and Harry) with **orders** and **customerPayments**. Sue and Harry send orders to the **Supplier** for more stock – **stockOrder**, and the **Supplier** sends them a **supplierInvoice**. Sue and Harry confirm delivery – **confirmationOfDelivery**, and send a **supplierPayment**.

Expanding processes

The level 1 data flow diagram in Figure 4.5 identifies the major system processes and data flows between them. The single process on the context diagram is expanded to partition the system into its main constituents, each part being an identifiable function or group of functions. The Just a Line system has been split into three main areas of activity: activities

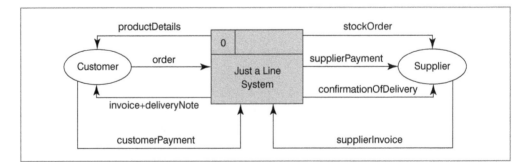

Figure 4.4 Context diagram for Just a Line

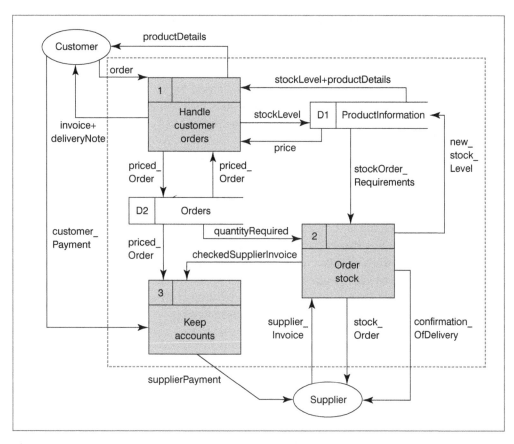

Figure 4.5 Just a Line: level 1 diagram of the current system

concerned with handling customer orders, activities concerned with ordering from the supplier and activities concerned with keeping the accounts. These three areas are logically distinct.

The external entities need only be shown on the context diagram; they may be shown at lower levels if this adds to the clarity of the diagram. External entities are shown here at levels 0 and 1, but not at level 2 (see below).

Each of the level 1 processes can in turn be expanded or decomposed into a level 2 data flow diagram to reveal more detail. For example, the diagram in Figure 4.6 expands process 1 'Handle customer orders'. This diagram shows that processing a customer order involves three stages:

1. agreeing the order provided that there is sufficient stock to meet it
2. pricing the order and filing it
3. making up the order and dispatching it with the invoice and delivery note.

There are three further points of interest, as follows.

1. On this diagram the data stores are shown crossing the boundary. This indicates that these data stores have already appeared on a higher-level diagram. In this situation it is also permissible to show the data stores completely outside the boundary.

2. The asterisk (and accompanying diagonal line) in the bottom right-hand corner of a process box (see Figure 4.6) denote that this is a bottom-level process, i.e. one that is not further decomposed. This is a useful notation because it indicates that there will be a process definition (see Section 4.4) for this process.

3. Just as the outline of the single process box on the context diagram becomes the boundary of the level 1 diagram, the boundary of each lower-level data flow diagram is provided by the frame of the higher-level process it expands (see Figure 4.7).

Processes can continue to be expanded in this way until the required level of detail is reached. Not all processes have to be decomposed to the same level. A simple process may not be decomposed further than level 1; a more complex process may be expanded to several levels.

When has the required level of detail been reached? As with many other aspects of modelling, understanding this is something that comes with experience. One simple rule of thumb is that decomposition should continue until each process has no more than one or two input data flows and one or two output data flows. Another guideline is that each process should represent a single task. A precise description of what happens in each of the bottom-level processes is given in the process definitions; if this description is more than half a page in length, the process probably needs to be decomposed further.

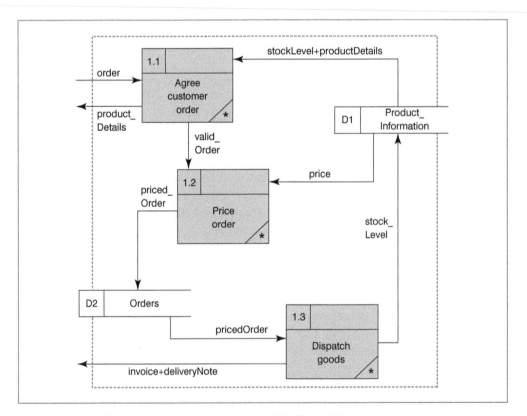

Figure 4.6 Just a Line: level 2 expansion of process 1 in Figure 4.5

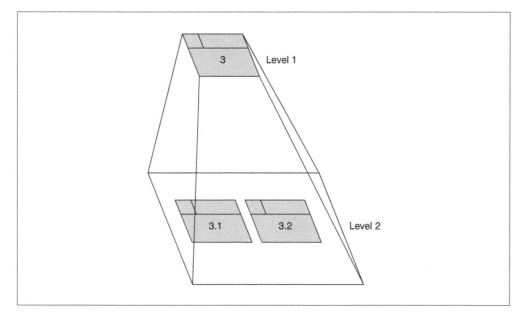

Figure 4.7 Relationship of process 3 at level 1 to its sub-processes at level 2

Numbering

On the level 1 diagram, processes are numbered 1, 2, 3 ... (see Figure 4.5). If a process at any level is decomposed, the sub-process numbers are prefixed with the number of the process they decompose. For example, if process 3 at level 1 is decomposed, the processes at level 2 will be numbered 3.1, 3.2 ... (see Figure 4.7). Each process, therefore, has a unique number, which is used for cross-referencing between the various models of the system. For example, a process will be referred to by this number when described in its corresponding process definition. A detailed data dictionary may also refer to process numbers.

It is tempting to make the process numbers represent the order in which the processing is done. However, this temptation should be resisted; the numbering is for reference purposes only. In order to avoid cramming too much information into one diagram, data flow diagrams make no attempt to model the processing sequence.

Labelling

In this text we have adopted a standardized notation for labelling. Labels for data flows are written in lower case; where a label consists of more than one word, the words are run together with no spaces but with a capital letter at the start of the second and any subsequent new words – for example, customer name becomes **customerName**. Labels for data stores follow the same convention, except that the first word is capitalized, e.g. **ProductInformation**. External entities are labelled in the same way as data stores.

The labels on the data flow diagram are as important as the numbers. Data flow diagrams see the system from the point of view of the data moving through it – this should be reflected in the choice of descriptions for processes and names used to label data flows, as noted below.

■ Process labels should be short sentences describing what the process does, e.g. **Price order** and **Order stock**. Normally they contain an active verb.

- Data flows should be labelled with nouns describing the data flowing along them, e.g. **order** and **customerPayment**. If you are tempted to label a flow with an active verb, e.g. **displayPrice**, it probably means you are trying to make a data flow do the work of a process.
- Data stores should be labelled to indicate the type of data stored in them. Each data store has a unique reference number. To distinguish them from processes, data stores are often numbered DI, D2, etc., or MI, M2, etc. for manual data stores. The prefix D is used where no reference is made to how the data store is implemented.
- Labels should carry as much meaning as possible, but should be short, to avoid cluttering the data flow diagram. Data flows and the contents of the data stores should be documented in a data dictionary (see Chapter 5).

Conventions of data flow diagrams

At first glance, data flow diagrams may look as if they are drawn and labelled very informally, but when used correctly the diagrams have a precise meaning. The technique of drawing data flow diagrams is underpinned by a set of '**syntax**' rules or conventions, which give them rigour and allow them to take their place in the total picture of the system provided by the combined set of structured modelling techniques. If these rules are not adhered to, the diagrams are almost meaningless. A system developer should regard these rules with the same degree of attention as a programmer whose syntax will be checked by a compiler. In fact, if a CASE tool is used (see Chapter 11), it will not allow diagrams to flout the rules.

Balancing rules

The balancing rules require that the data flows entering or leaving a parent diagram must be equivalent to those on the child diagram. The context diagram of the Just a Line system (Figure 4.4) has exactly the same data flows as those crossing the boundary of its child diagram, the level 1 data flow diagram (Figure 4.5). New external flows may not be introduced at lower, more detailed, levels. In Figure 4.8 the balancing rule has been broken twice in the decomposition of process 2:

- a flow S, which was not on the level 1 diagram, has appeared crossing the boundary at level 2
- the flow M appears on the parent diagram but not on the child diagram.

The flows P and Q are introduced in the child diagram, but as these are internal to the child diagram they do not infringe the balancing rule.

Modelling data stores

In the interests of keeping clutter on the data flow diagrams to a minimum, data stores that are local to a process need not be drawn until the process is expanded in a lower-level diagram. For example, in the parent diagram in Figure 4.8, no data stores are modelled. The decomposition of process 3 (Figure 4.9) reveals a data store local to process 3. At level 1 the data store DI is not used by any process except process 3, i.e. it is not shared by the processes at level 1. At level 2 the data store is used by more than one process and therefore must appear in the diagram.

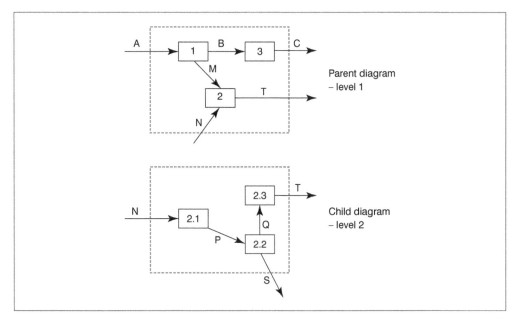

Figure 4.8 Parent and child diagrams do not balance

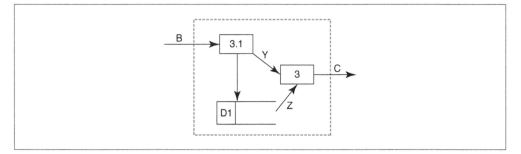

Figure 4.9 Decomposition of process 3 from level 1 in Figure 4.8, revealing a local data store

Expanding data flows

The balancing rule prohibits the uncontrolled introduction of data flows at lower levels. However, if all the lower-level flows were to appear on the context and level 1 diagrams, they could get very cluttered. To avoid this, it is permissible to bundle flows using a unifying label at the top level. This unifying label (for a bundled data flow) can then be unravelled (or decomposed) as required at the lower levels. Figure 4.10 shows the data flow **amendment** at level 1, which decomposes at level 2 into two separate flows, **cancellation** and **alteration**. Decomposition of data flows must be documented in the data dictionary:

$$amendment = cancellation + allteration$$

Data dictionary notation is described in Chapter 5.

Flows to and from data stores

All flows should be labelled except when a flow to or from a data store consists of exactly the same items of data as are in the data store: the flow represents the storing or extraction of a

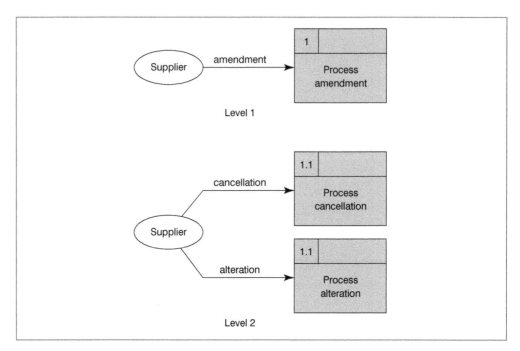

Figure 4.10 Decomposing a data flow

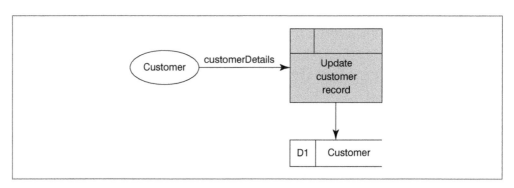

Figure 4.11 Unlabelled flow

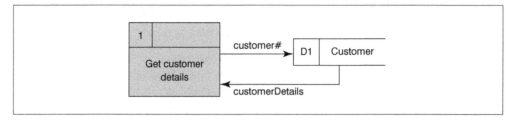

Figure 4.12 The flow **customer#** should not appear on the diagram

whole record. Figure 4.11 shows a fragment of a data flow diagram where the data store Customer, containing customer details, is updated by an unlabelled flow whose content is the same as that of the data store, i.e. a whole customer record has been added. Conventionally, a flow into a data store means that the data store is being updated in some way. Selection criteria used to find and extract data are not modelled on data flow diagrams. It is tempting, for example, to use the technique illustrated in Figure 4.12 to indicate that **customer#**[2] is used to find a particular record in the data store Customer. However, this temptation should be resisted – the flow customer# should not appear on the diagram. The justification, again, is that the diagrams must be kept as uncluttered as possible; selection criteria can be assumed.

Modelling deletion of a record from a data store

Sometimes we need, specifically, to model the deletion of a record from a data store, e.g. an old stock record from the stock file or a customer record from the customer file. This can be modelled as shown in Figure 4.13. However, for the sake of keeping a diagram uncluttered, deletion of records is often assumed as part of a general update process (see Figure 4.14).

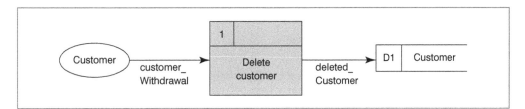

Figure 4.13 Modelling record deletion 1

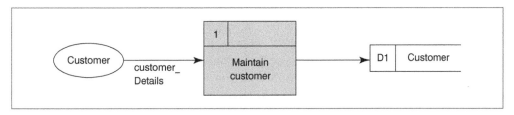

Figure 4.14 Modelling record deletion 2

Crossed data flows

Data flow diagrams drawn with crossing data flows can be hard to read. Crossing flows should be avoided by rearranging the elements of the diagram or duplicating data stores or external entities.

Different stages

Data flow diagrams can be used to model the system at different stages of development. They can be used to model both the existing and the new systems. A data flow diagram of an existing system can be used to model either *how* the existing system works or *what* the existing system does. A DFD of an existing system that focuses on the mechanics of the current operation of the

[2] The symbol # means number.

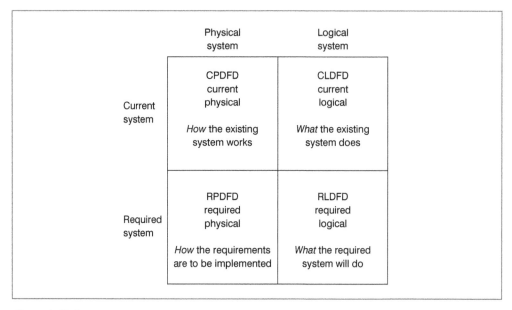

	Physical system	Logical system
Current system	CPDFD current physical *How* the existing system works	CLDFD current logical *What* the existing system does
Required system	RPDFD required physical *How* the requirements are to be implemented	RLDFD required logical *What* the required system will do

Figure 4.15 Stages where a data flow diagram may be used

system, i.e. how the existing system works, is known as a ***current physical data flow diagram (CPDFD)***. A data flow diagram of the existing system that focuses on the logic of the system – what the existing system does, without reference to how it does it, is known as a ***current logical data flow diagram (CLDFD)***. In the same way, data flow diagrams of the new (required) system can focus on either the logic (what it will do) or the implementation (how it will work).

Figure 4.15 summarizes the stages a data flow diagram can model. If all stages are modelled, the order will be current physical DFD, then current logical DFD, then required logical DFD and finally required physical DFD. However, a system developer may not find it necessary or desirable to model all the stages. There may be no current system, in which case the ***required logical data flow diagram (RLDFD)*** will be the starting point. If the current system is complex and used by many people, accounts of how it works may be different and even contradictory. In this case a CPDFD may be useful for the system developer both to sort out his or her own ideas and as a focus for discussions with the clients to resolve conflicts. Some system developers, however, prefer not to draw CPDFDs as they feel it prevents them from producing a radically different (and better) design – once they have modelled the system one way, it is hard to rethink the basic structure. A detailed discussion of the differences between physical and logical data flow diagrams is given in Section 4.3.

All DFDs should be labelled with the name of the system, the level and the stage, e.g. Just a Line, level 1 CPDFD.

Limitations of data flow diagrams

Data flow diagrams purposely limit the information they show. They make no attempt to model the following features:

■ the sequence in which processes occur
■ the time intervals at which processes occur

- detail about the structure of data flows or data stores (what does an **order** consist of, what data is stored in **ProductInformation**?)
- how often a process is repeated
- conditions governing the occurrence of certain events (e.g. if the card design is in stock, then the order is accepted; or if too much money is tendered then change is given); if a flow ever happens, it should be modelled.

If all this information were recorded on data flow diagrams, they would become too cluttered to be useful.

Checking data flow diagrams

We can never prove that a data flow diagram is correct – we can never know for certain that we have drawn what really exists or what the clients really want. However, data flow diagrams

---

Data flows

1. Data flows can take place between two processes, an external entity and a process, a data store and a process, but not between external entities and data stores, data stores and data stores, or between two external entities.
2. A data flow label should change if the content of the data flow changes or if significant processing is done to it. If two data flows have exactly the same label it means that the content is identical.
3. All flows should be labelled except where a flow going into or coming out of a data store consists of exactly the same data items as the data store.
4. A data flow into a data store implies data is going into that data store, i.e. it is being updated in some way. Selection criteria used to extract data are not modelled.
5. A flow cannot be made to model an action – the flow label is just a name for the bits of data that the flow contains, e.g. 'order'.
6. Data flows should not be allowed to cross – duplicate external entities or data stores should be used.
7. No new external entities or I/O flows (flows to and from external entities) can be introduced in lowern level diagrams – they must all be introduced in the context diagram. Flows may be 'bundled' at higher levels and decomposed in lower-level diagrams.
8. If a flow ever occurs, it is modelled. The data flow diagram should not attempt to model the conditions under which a flow occurs.

Processes and data stores

9. Each process must have a unique reference number.
10. Each data store must have a unique number and name.
11. Duplicated external entities and data stores must have 'cut-off corners'.
12. If a data store is shared between two processes at any given level, it must be modelled. If, at a given level, a data store is internal to a process (i.e. only used by that process), it need not be modelled.

Overall diagram

13. Every data flow diagram must have a boundary marked on it.
14. All diagrams must be labelled, e.g. Just a Line: Context diagram.
15. Sequence and timing of flows and processes relative to each other are not modelled on a data flow diagram – the data flow diagram has nothing to say about the order in which flows or processes occur.
16. With levelled data flow diagrams the boundary at any given level is the perimeter of the process box it decomposes. Flows must balance between levels.

Figure 4.16 'Syntax' rules for data flow diagrams

can be checked for internal completeness and consistency. The basic points to check are listed below.

- Does each process receive all the data it requires? A process that prices orders (Figure 4.6) will need the relevant prices as well as details of what has been ordered.
- Does any data store have only data flows coming out and nothing going in? Data must be updated somewhere: suppliers change their products and prices, customers change their addresses. The system must allow for this.
- Does any data store appear to have only incoming data that is never used, i.e. no data flow out of the data store? Is this correct?
- Are the data flow diagrams consistent across levels, i.e. have the balancing rules been infringed?
- Do all the external entities appear on the context diagram? Are any introduced at lower levels?
- If a named bundle of flows is decomposed at a lower level, is this documented in the data dictionary?
- Do all the flows that should have a label have one? Are the labels documented in the data dictionary?
- Do all the flows on the diagram represent genuine flows of data? Do any of them represent flow of control? If the diagram reads 'after process A you do process B', it is showing flow of control. The numbers on the process boxes are for reference only; they do not represent the order in which the processes should be done. The arrows represent data flows only, they do not show which process should be done next.

Diagrams should be checked carefully to see that they do not infringe any of the data flow diagram 'syntax' rules. A summary of the 'syntax' rules for data flow diagrams is given in Figure 4.16.

4.3 Data flow diagrams: part 2
Logical and physical data flow diagrams

A *physical* **data flow diagram** models how the system does or will operate. Current physical data flow diagrams of an existing system describe the system in the user's terms – symbols on the diagram correspond to physical objects the user can recognize. Data flows may represent identifiable bits of paper; a data store may represent an identifiable ledger book or filing cabinet. In the Just a Line system, the data stores should represent objects mentioned in the interview: the card store, the filing cabinet, the price list and the design list (see Figure 4.17). The data flows **supplierInvoice** and **pricedOrder** correspond to identifiable documents. The diagram records other relevant physical details, e.g. that orders come in *by telephone*, that orders are filed *in triplicate* and a *signed copy* of the supplier invoice is returned to the supplier as confirmation of delivery. A physical data flow diagram may specify *who* does a particular job or *where* it is done; it may also record the order in which data is stored in a data store (see the data store, M2 **Orders**, in Figure 4.17). Current physical DFDs show the 'what, how, who, when and where' of a system (Robinson and Prior, 1995).

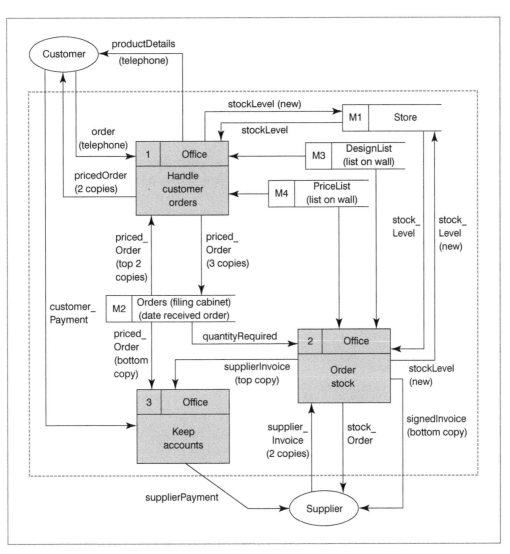

Figure 4.17 Just a Line: level 1 CPDFD

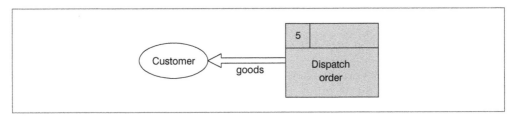

Figure 4.18 Modelling material flow

Sometimes it is useful to model material (or physical resource) flows rather than just data flows. We might want a diagram to emphasize the movement of physical objects of interest. This is done using broad arrows, as shown in Figure 4.18. The main characteristics of current physical data flow diagrams are summarized in Figure 4.19.

A **required physical data flow diagram** (**RPDFD**) describes how the new system will work and may indicate such things as which parts of the system will be automated and which will remain manual, what physical devices will be used, and what input and output forms or reports will be used.

In contrast, a *logical data flow diagram* reflects what a system does or will do without reference to how it does or will work. It separates the essential **functionality** of the system (what it does) from the implementation detail (how it does it). This separation of concerns is an essential step for the system developer as the new system usually has to do everything that the existing system does, plus meet the new requirements and solve any current problems. What the existing system does, therefore, will filter through to the new system, although how it does it will almost certainly change in the new implementation. The current logical DFD retains *what* happens and removes the *how*, *who*, *when* and *where*. Figure 4.20 summarizes the steps from the physical to logical DFD.

Current physical DFDs

1. Show how the existing system works in implementation terms, i.e. show physical details about how the system currently works. They show:
   - who does a particular job
   - where a particular job is done; if it is relevant indicate when it is done
   - how a flow is communicated (by telephone, post, email, internal mail, visual display)
   - how data is stored (in a loose-leaf file, on a whiteboard, in a cabinet, in a shoe box)
   - the sequence in which data is stored (date order, customer name order)
   - the nature of the document (signed copy of invoice, annotated price list)
   - material flows

2. Model the system in the user's terms, i.e. on the data flow diagram describe elements of the system in terms that the user recognizes.
   - Describe data stores in terms of the user's descriptions even if these are obscure to an outsider. If the user calls a list of products his 'Bible', have a data store labelled 'Bible' on the diagram.
   - Label data flows in terms that the user will identify with: if they know a form as the 'pink order', or the 'LC1', use these terms on your diagram.
   - Users often describe a system in terms of its functionality, i.e. in terms of the different jobs that are done. Your diagram should reflect their view of the system – their view of the way it breaks down into separate jobs. This may dictate the entire decomposition of the diagram.

3. Model the system exactly how it is without any attempt to make it look more logical or improve it in any way.
   - As you draw and document the current physical system, you may become aware of several anomalies such as the same data being stored in more than one place, jobs being done in an inefficient sequence, etc.
   - Improvements can be incorporated in the current logical model; a good current physical model shows things exactly as they are, warts and all.

Figure 4.19 Main characteristics of current physical DFDs

To move from current physical to current logical DFD

1. Get rid of references to physical implementation.
   - Remove references to who, when, where and how a job is done or data is stored.
   - Get rid of material or resource flows and concentrate on the associated flow of data or information only.
   - Rename data stores and data flows so that the name reflects the content; this is the time to get rid of idiosyncratic user names or very physical names.
   - Remove processes that are required purely to support the current implementation, e.g. those that reflect the way someone does their job at the moment but that could be done otherwise.

2. Make minor 'improvements' in logic.
   - If several data stores in the current implementation store essentially the same data, these could be combined.
   - If a single data store in the current implementation stores data about several things that are logically distinct, consider creating several separate data stores.
   - Remove any data stores that are used only as an implementation-dependent time delay between processes. For example, if Sue kept orders in a shoe box until she had time to price them, the shoe box might appear on the current physical DFD but would disappear in the current logical DFD.

Figure 4.20 Moving from current physical to current logical DFD

Comparison of current physical DFD and current logical DFD for Just a Line

The difference between physical and logical data flow diagrams can be illustrated from the Just a Line case study by comparing Figure 4.17 (reproduced for convenience as Figure 4.21) with Figure 4.22.

- On the current physical DFD we record that orders come in *by telephone*, that there are *two copies* of the **supplierInvoice** and *three copies* of the **pricedOrder**; there are references to the *top copy of the* **supplierInvoice** and the *bottom copy* of the **pricedOrder**. Details about the current implementation are omitted from the logical DFD.
- The current physical DFD records that all three processes are done *in the office*. This information about how the current system is implemented is discarded in the current logical DFD.
- The data stores on the current physical DFD represent actual physical things in the current system: the **Store** where the supplies of cards are stored, and the **DesignList** and **PriceList**, which refer to the two lists pinned up near the telephone. The current logical DFD extracts the information that will be used in the new system, i.e. the supplier's name and address, product codes and prices from the **PriceList**, the product descriptions from the **DesignList** and the amounts in stock from the **Store**. This data is combined in a new data store **ProductInformation**. The new system will certainly need to store this information, although in a different form from the way it is recorded in the current system. The relevant information is therefore recorded in the current logical DFD but the old method of storing it (in the **Store** and on the **DesignList** and **PriceList**) is discarded.

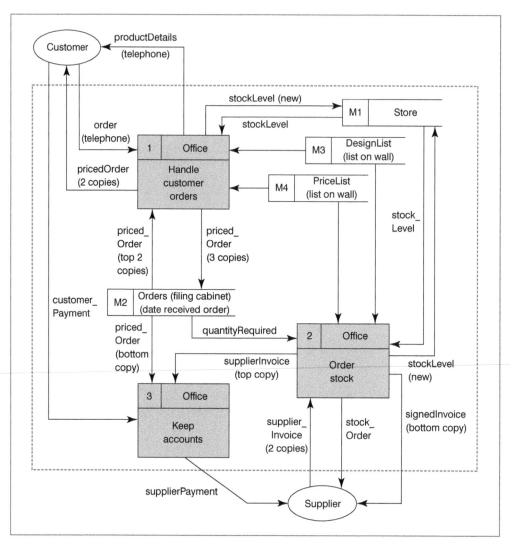

Figure 4.21 Just a Line: level 1 CPDFD

- Information about the order in which data is stored in the current system is discarded, i.e. that orders are stored in the order in which they are received.
- The data stores on the current logical DFD have a reference number beginning with D rather than M – they no longer refer to manual data stores.
- The two copies of **pricedOrder** sent to the customer with his or her order are renamed **invoice+deliveryNote** on the current logical DFD. This tells us more about the purpose of these documents and less about which physical bits of paper are sent.
- The only other thing that is different is that the bottom copy of the **supplierInvoice**, which is signed and returned to the **Supplier**, becomes **confirmationOfDelivery** on the current logical DFD. This tells us more about its function and less about how it is done physically.

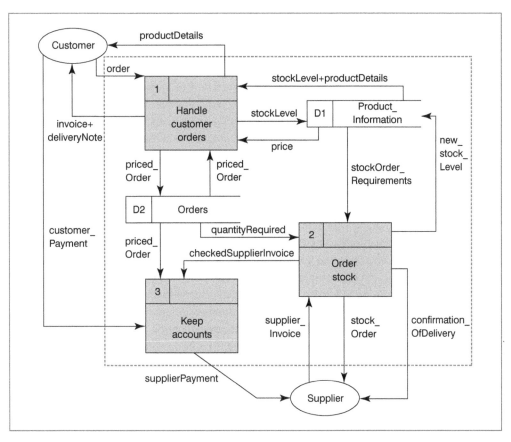

Figure 4.22 Just a Line: level 1 CLDFD

Required logical DFDs

The new system must do everything that the old one did plus solve any identified problems and meet any new requirements. New requirements will be outlined in the problem definition and requirements specification. The simplest approach to developing the required logical DFDs is to combine the current logical DFDs with the list of requirements outlined in the requirements specification, i.e. add the requirements for the new system to the current logical DFD.

Another approach to developing the required logical model is to start by identifying all the system outputs (from the current logical DFDs plus the requirements specification), drawing the outputs on the required logical DFD, documenting them in the data dictionary and working backwards. For example, if we have to produce an itemized bill, where do we store the data we need to print on the bill and, working back from there, where does that data come into the system?

Another school of thought requires us to find all the bottom-level processes on the CLDFD with their inputs, outputs and data stores, list them in a completely unstructured way or even draw them on a huge sheet of paper, sort out which processes 'go together' and structure them into a levelled set of DFDs. This is a bottom-up approach designed to help the developer get completely away from the decomposition of the old system.

We will adopt the first approach and combine the current logical DFDs with the requirements for the new system documented in the requirements specification.

Just a Line requirements for the new system

The new system for Just a Line must incorporate several new requirements. The system developer will already have listed most of these in the requirements specification (see Chapter 3). The list will almost certainly be added to during the investigation of the current system, often because users will raise problems, which for various reasons they did not mention earlier. Also, system developers, from past experience, will spot features they think should be incorporated in the new system.

The list of requirements includes some points not mentioned in the original interview and that we will assume were mentioned at supplementary interviews. Harry has decided that he does not want the accounting side of the system computerized at this stage. He feels that he has a nice simple system in place and does not want to change it. This has been agreed with the system developer, who feels that if Harry changes his mind later it will be possible to buy an accounting package and incorporate that into the computerized part of the system. Sue, on reflection, has also stipulated that there is a part of the system she does not want computerized at this stage. She feels that it would be more effort than it is worth to keep track of stock levels on the computer and fully automate the stock control. The plan is to continue to handle stock control manually, but to have a list of items on pending customer orders. The list is to be produced by the computer to help her with supplier ordering. An extract from the specification of requirements is given below.

Just a Line requirements list

■ The system must produce a personalized combined price list and order form to be sent out to existing customers. A non-personalized version of the same form will be distributed for publicity purposes.

■ To facilitate the above requirement, the system will maintain a file of customer details. Customer numbers will automatically be assigned to new customers.

■ The system must check whether a customer order is valid by checking that the card design ordered is one that is stocked. Checking that there is enough of that particular design in stock to meet the order will remain a manual process for the moment. The computerized system must keep a record of pending customer orders (not yet made up or delivered).

■ When the orders are made up the system must produce invoices to go with the goods.

■ The process of preparing a supplier order will remain manual but the system must be able to list products and quantities on the current batch of pending customer orders, to facilitate supplier ordering.

Sue and Harry feel that the combined price list + order form will solve most of their current problems with ordering. Customers will either send in the order form or, if they still telephone in the order, at least they will have a price list in front of them, which will avoid the tedious process of reading over the telephone the list of items available. Pricing orders and producing the invoices will be automated. While the main stock control and ordering from suppliers is still manual, Sue will be helped when ordering by the list of items required to fulfil current pending

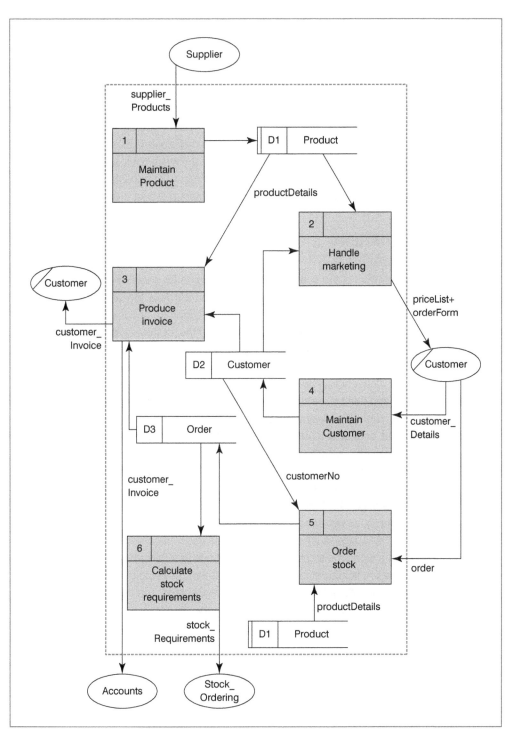

Figure 4.23 Just a Line: level 1 RLDFD

customer orders. These requirements are modelled in the level 1 required logical data flow diagram for the Just a Line system, shown in Figure 4.23.

Notice that **Keep accounts** and **Order stock**, which appeared as processes on the data flow diagrams of the current system, are now shown as the external entities **Accounts** and **StockOrdering**. As the clients have requested that these functions remain unaltered, they are considered to be outside the current domain of interest and therefore external to the system. However, they still interface with the system as they receive system outputs. The system will keep a file of information about products, which will include descriptions of products stocked and their prices. This file must be kept up to date. However, the system will not attempt to keep track of current stock levels. The system will also maintain a file of customers to facilitate the production of the price list + order form, and a file of current and past orders.

4.4 Process definitions
What are process definitions?

Process definitions form part of the supporting documentation for data flow diagrams; they describe in detail what is happening in the process boxes on the data flow diagram.

A data flow diagram at any level may have both processes that are decomposed at lower levels and processes that are not decomposed further. Only the 'bottom-level' or 'elementary' processes are described in process definitions. High-level process boxes simply serve as structuring devices; they bundle together lower-level processes to provide a convenient overview of the system. All the action takes place in the elementary processes; therefore these are the only processes that need to be described. Elementary processes should be marked by an asterisk in the bottom right-hand corner of the process box on the DFD (see Figure 4.6).

An elementary process on a DFD already has a short description such as 'Produce invoice' or 'Handle marketing' (see Figure 4.23). Such labels are sufficient for the purposes of the diagram – they give us an overview of what is happening in the system. However, as the development of the system progresses we need to specify more precisely what happens in the processes; ultimately we need to tell the programmers what the processes are doing.

Before structured techniques were used to develop systems, developers described functionality in a natural language, such as English. These descriptions often proved to be ambiguous, incomplete and inconsistent. Process definitions use descriptive techniques that avoid the ambiguity inherent in natural-language descriptions; these include *structured English*, *decision tables* and *decision trees*. To increase precision and consistency, nouns used in process definitions are defined in the data dictionary.

Structured English

Structured English is a limited and structured sub-set of natural language, with a syntax similar to a block-structured programming language. The following set of constructs is suggested.

- A sequence construct – statement 1 before statement 2, for example:
    1. Calculate total order cost
    2. Add delivery charge
- Two decision constructs, for example:
    if ... then ... else construct
    case construct, for deciding between more than two alternatives

■ One or two repetition constructs, for example:
while ... do
repeat ... until

Guidelines for writing structured English

■ Develop a consistent style. Once you have selected a set of constructs (such as the above) stick to these; do not introduce new ones.
■ Use layout conventions to help reveal structure and meaning in the same way as they are used in a high-level programming language. Look at the layout of the Just a Line example below to see how this works.
■ Use the data dictionary. Refer to nouns by their data dictionary names. If a word is not already in the data dictionary, add it.

Just a Line example

In a supplementary interview, Sue told the system developer how they decided whether or not a customer was entitled to free delivery of their order. This is how Sue described the process:

> To be entitled to free delivery, customers must place an order worth £30 and live within a five-mile radius or have been with us for more than a year. If customers do not live locally we post the cards.

This description has the drawbacks of many natural-language descriptions in that it is:

■ **ambiguous** – does it mean '£30 and (live within five miles or one year's custom)' or '(£30 and live within five miles) or one year's custom'?
■ **imprecise** – what is a 'five-mile radius' – is it five miles as the crow flies or five miles on the mileometer? What does 'local' mean?
■ **incomplete** – we assume it is '£30 or more'; that seems obvious in this case, but we should not have to make any assumptions.

Structured English

The system developer rewrote Sue's description in structured English, avoiding the pitfalls of the natural-language description:

```
if not local customer (*lives outside 20 miles radius*)
        charge P&P
else (*local customer*)
        if totalCost (*of order*) < £30 then
                set deliveryCharge
        else (*totalCost > = £30*)
                if distance < = 5 miles then
                        no deliveryCharge
                else (*distance > 5 miles*)
                        if customerTradeRecord > = 1 year then
                                no deliveryCharge
                        else (*customerTradeRecord < 1 year*)
                                set deliveryCharge
```

Comments are made between an opening parenthesis and asterisk and a closing asterisk and parenthesis: (* ... *). Terms with special meaning for the Just a Line system (customer, order, deliveryCharge, etc.) must be defined in the data dictionary. The Just a Line current logical data dictionary, therefore, must be updated as necessary to record the nouns introduced in the above structured English description:

```
customerTradeRecord
deliveryCharge
distance
P&P
totalCost
```

As the nouns are being used to describe the current system it is sufficient to describe them with an informal comment:

```
customerTradeRecord = *deliveryDate of first order*
deliveryCharge = *calculated at a flat rate per mile*
distance = *number of miles on mileometer*
P&P = *package and posting; cost of postal stamps plus flat rate
        for package*
totalCost = *total cost of order before adding delivery or postal
              charges*
```

Structured English is a useful technique for communicating with programmers as it expresses the logic of processes in a way that maps naturally on to code.

Decision trees

System developers use two other techniques for expressing process definitions: decision trees and decision tables. The free delivery process is expressed as a decision tree in Figure 4.24. The conditions that affect the outcome are listed along the top, the different possible outcomes (or actions) are listed in the right-hand column. The decision tree expresses the logic of the process clearly and graphically and has a visual immediacy, which makes it a useful technique for communication with a client.

Decision tables

Where a decision process involves a complicated combination of conditions and actions it may be more appropriate to use a decision table. The general format of a decision table is shown in Figure 4.25. A decision table is created by listing all the relevant conditions in the top left-hand quarter of the decision table and all the relevant actions in the bottom left-hand quarter of the decision table. The top right-hand corner of the table tabulates, in separate columns, all possible combinations of conditions. Each column, recording a possible combination of conditions, is called a rule. Below each rule, in the bottom right-hand corner of the table, the appropriate actions are specified. The conditions that affect the free delivery process are:

- local (*customer*)
- totalCost (*order*) < £30
- distance > 5 miles
- customerTradeRecord < 1 year

The possible actions are:

- charge for delivery
- charge P&P
- no charge

The rules dictating which combination of conditions results in which actions are shown in Figure 4.26.

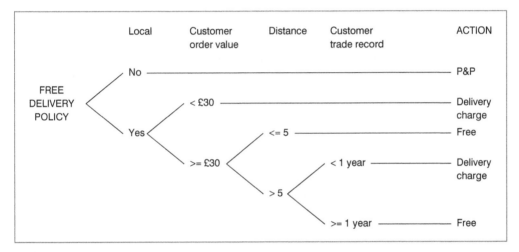

Figure 4.24 Free delivery process as a decision tree

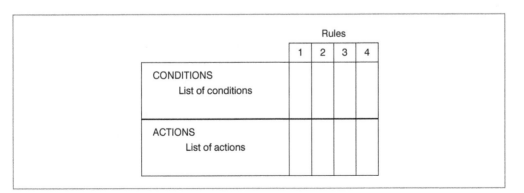

Figure 4.25 Decision table structure

		Rules				
		1	2	3	4	5
Local (*Customer*)		N	Y	Y	Y	Y
Total cost (*Order*) < £30		–	Y	N	N	N
Distance > 5 miles		–	–	N	Y	Y
Customer trade record < 1 year		–	–	–	Y	N
Charge for delivery			X		X	
Charge P&P		X				
No charge				X		X

Figure 4.26 Free delivery process as a decision table

Summary

Data flow diagrams model the flow of data through the system. Data flow diagrams are used to describe the system at different stages of development. They are mostly used in the requirements engineering stage to describe the current system and the required system. They are also used to model the system during the feasibility study and at the design stage. Data flow diagrams are used to force structure and coherence on the mass of facts gathered during the fact-finding stage of requirements engineering. They give a semi-technical, pictorial representation of the system, which will form the basis of good communication with the client. In the process of drawing the data flow diagrams and discussing them with the client, inconsistencies, omissions and misunderstandings may be observed and corrected. Data flow diagrams use decomposition and abstraction to limit the amount of detail tackled at any one time. They use abstraction in that they attempt to model only certain limited aspects of the system: they deliberately make no attempt to show the sequence or frequency of events, nor the circumstances under which a flow will occur. These and many other features of the system are modelled by other techniques. DFDs use decomposition in that they partition the system, in a top-down manner, into its major processes, allowing us to concentrate on one part of the problem at a time. Each major process may in turn be partitioned. In this way the system can be viewed as a whole, or selectively in detail.

Process definitions describe what happens in the bottom-level process boxes of a set of data flow diagrams. Natural language descriptions can be ambiguous, incomplete and inconsistent. Process definitions are normally expressed in structured English or by a decision table or a decision tree. Decision tables are useful for describing complex processes with many interdependent conditions and many possible actions. Decision trees have the advantage of immediate at-a-glance clarity but can be inflexible. Structured English is very flexible, makes use of the data dictionary and is much favoured by those with a programming background as it leads naturally, through *pseudocode*, to many high-level programming languages.

Exercises and topics for discussion
What can you remember?

You will find all the answers in the chapter

Revision questions on modelling and data flow diagrams: part 1

a) Define the terms *abstraction* and *decomposition*.

b) What perspective of the system is given by the following techniques?
   (i) data flow diagrams
   (ii) process definitions
   (iii) data dictionary
   (iv) data model
   (v) entity life histories

c) At what stages in the development of a software system are data flow diagrams used?

d) What are the five basic elements of a data flow diagram? Explain the purpose of each of these elements.

e) Why do we need to use levelled sets of data flow diagrams?

f) Which parts of the system are modelled by the context diagram?

g) On which levels of a data flow diagram is it permissible to draw external entities?

h) What does it mean when a data store is shown crossing the boundary of a data flow diagram?

i) What is signified by placing an asterisk in the bottom right-hand corner of a process box?

j) How can you tell when a data flow diagram has been sufficiently decomposed?

k) If process 3.1 of a level 2 data flow diagram is decomposed into four processes at level 3, what will they be numbered?

l) Why are some data stores given a reference number starting with M and some with D?

m) What is meant by the term bundled data flows and how do they work?

n) When is it permissible to have an unlabelled data flow?

o) Summarize the main differences between physical and logical data flow diagrams and between current and required data flow diagrams. In what order are they drawn?

p) Data flow diagrams purposely limit the information they attempt to show. What information do data flow diagrams deliberately omit?

q) Give a list of points you should check on a data flow diagram.

Revision questions on data flow diagrams: part 2

r) What symbol is used to model the movement of physical objects of interest on a CPDFD?

s) List the four aspects of a current physical data flow diagram (three of which begin with the letter w) that are removed when we draw a current logical data flow diagram. What do we retain?

t) Identify three different methods of getting from the current logical data flow diagram to the required logical data flow diagram.

Revision questions on process definitions

u) What is the purpose of a process definition; what does it describe?

v) What advantage does a structured process definition have compared to a description in a natural language such as English?

w) Name three techniques used to write process definitions.

x) Suggest a set of language constructs that would be useful for writing process definitions in structured English.

y) What are the advantages of describing a process by using the following techniques?
  (i) a decision tree
  (ii) structured English

z) When would it be appropriate to use a decision table to describe a process?

Exercises on data flow diagrams: part 1

4.1 Study the level 2 current logical data flow diagram of the Just a Line system in Figure 4.6. Check that it is consistent with the level 1 diagram in Figure 4.5 – is the balancing rule infringed?

4.2 In Figure 4.8 the parent and child diagrams do not balance. Redraw the child diagram so that the flows balance with the parent diagram.

4.3 *Automatic ticket machine*. The management of the Metro train service has requested an automatic ticket machine to function as follows. The user of the machine asks for a destination and ticket type (e.g. single, return or day return) and the machine displays the price. When enough money has been put in, the machine issues a date-stamped ticket containing the issuing station, the destination, ticket type and price. If appropriate, the machine also gives change. The machine records the number of tickets issued for each destination and the number of tickets of each type for each day. Once a week these statistics are transferred to the main computer.

  (a) Draw the context diagram of the automatic ticket machine system.
  (b) Draw a level 1 CLDFD of the automatic ticket machine system.

4.4 *Capital Taxis*. Capital Taxis is a firm that provides transport for passengers based in London and the south-east. Organization of these journeys is handled centrally from Capital's control centre near Marble Arch.

  When a call is received in the control centre, the receptionist writes down the call details on a pre-printed form. The pick-up address is identified from a map-book, together with the map reference coordinates. On completion of the call the form is placed in a central collection point with other call forms.

  The distribution clerk collects the call forms from the central collection point and, from the details on the forms, decides which division should deal with it (the control centre is divided into four divisions – North, South, East and West). The clerk also tries to pick out any duplicate calls at this point. The call form is passed to the relevant division, where an assistant examines it together with location and availability information on the taxis in the division. The journey is allocated to the taxi that is nearest the pick-up address and currently free.

(a) Draw a context diagram of the current system at Capital Taxis.

(b) Draw a level 1 current physical data flow diagram of the system at Capital Taxis.

4.5 *Department store credit system*. A large department store has an arrangement whereby regular customers can pay for purchases using a store credit card. When a customer wishes to make a purchase, he or she presents a credit card to the assistant. The card is then passed through a machine to check that the card holder has sufficient credit. The sale is then authorized or refused by the company.

If the sale is authorized, the transaction is carried out using a two-part voucher. This records details of the customer and the transaction. The customer signs the voucher and then keeps one copy. The assistant puts the other copy of the voucher in the till. At the end of each day all vouchers are collected and sent to the accounts department.

Each credit card holder receives a monthly statement showing the details of payments and purchases since the last statement, any interest due, the total amount owing and the minimum amount that must be sent to the company. This is calculated as a percentage of the amount owing or £10, whichever is the greater. Records of customer transactions are archived and kept for at least five years.

(a) Draw a current physical context diagram of the system described above.

(b) Draw a level 1 current physical DFD of the system described above.

4.6 *Estate agent's system*. A potential buyer contacts the office and is given details that meet his requirements. He is then added to the mailing list and will be sent particulars of suitable properties as they come onto the market.

Sellers ask for a valuation and an agent is sent to value the property. Valuing the property consists of making notes of the property details, measuring rooms and estimating the price the property could fetch. Afterwards the sellers are sent a standard letter outlining the service provided by the firm and quoting a price for the property. Property detail leaflets are typed from the valuation notes. Property details for the newspaper have to be ready by Monday at 4pm.

Buyers put in an offer on a house to the agent, who conveys it to the seller. If the seller accepts the offer then the house is marked as 'under offer'.

(a) Draw a context diagram of the current physical system at the estate agent's office.

(b) Draw a level 1 data flow diagram of the current physical system at the estate agent's office.

4.7 *X-ray system*: an exercise in drawing levelled data flow diagrams.

1. *Overview – level 1 DFD*. A hospital X-ray clinic does X-rays by appointment. A patient is given an X-ray request form by a GP. The patient calls at the clinic to make an appointment and receives an appointment card.

2. When the appointment falls due, the patient reports to the clinic and presents the appointment card. The patient is X-rayed and a report is prepared, one copy of which goes to the GP.

3. *Expansion of paragraph 2 – level 2 DFD.* When the appointment falls due, the patient reports to the clinic and presents the appointment card. The appointment is checked by the clinic receptionist and the X-ray request retrieved. A history request slip is sent to the filing room so that the patient's reports and X-rays can be extracted and sent to Radiography. The X-ray is taken and clipped to any patient history. This information is passed to the consultant, who produces a new report. One copy is filed with the X-rays and old reports; another copy is sent to the GP.

4. *Checking understanding.* The analyst now feels she has a reasonable understanding of the system. To check, she 'walks through' the level 2 DFD with the people responsible for each stage. The consultant's secretary tells her: 'When the new report is ready, I clip it to the past X-rays and reports and return the updated history to the filing room, who copy it, file one copy and send the other copy to the GP.'

Using the information given in paragraphs 1 and 2, do the following.

(a) Draw a context diagram for the system at the X-ray clinic.

(b) Draw a level 1 diagram for the system at the X-ray clinic.

(c) Using the extra information given in paragraph 3, expand your level 1 diagram as appropriate to produce a level 2 diagram.

(d) Using the information given by the consultant's secretary in paragraph 4, revise your level 2 diagram to take account of the new information.

(e) Expand your level 2 diagram as appropriate to produce a level 3 diagram.

4.8 *Milk delivery system.* The level 2 data flow diagram in Figure 4.27 contains several 'syntax' errors. These errors should be apparent without further information about the system. List the errors. A correct version of this diagram is shown in Figure 4.30.

Exercises on data flow diagrams: part 2

4.9 Convert the level 2 current logical data flow diagram in Figure 4.6 to a current physical data flow diagram (CPDFD). Include the physical details supplied in the interview about the Just a Line system in Chapter 3 and those incorporated into the level 1 CPDFD in Figure 4.17.

4.10 The interview in Chapter 3 is intentionally brief. Drawing the current physical data flow diagram in answer to Exercise 4.9 should reveal some gaps in the picture of the Just a Line system. Make a list of questions you would ask at a supplementary interview.

4.11 Study the CLDFD for the Just a Line System in Figure 4.5. Write a brief description (as if you were explaining the diagram to your clients, Harry and Sue) of what happens in the system.

4.12 Use the information supplied in the Just a Line interview (Chapter 3) to expand the process 'Order stock' in Figure 4.5 to a level 2 diagram.

4.13 Discuss the relative merits of using physical or logical data flow diagrams to model a current system. What factors would influence your decision?

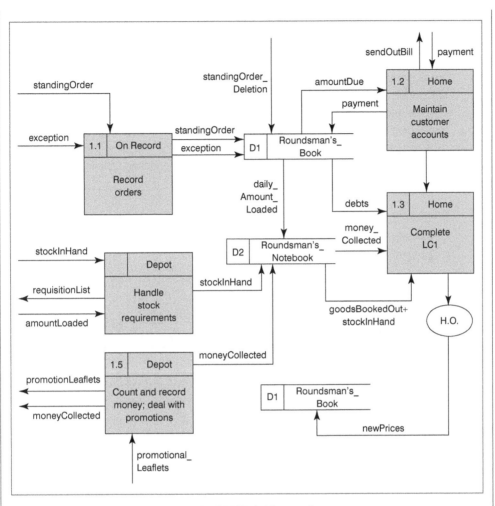

Figure 4.27 Milk delivery system: level 2 CPDFD (with errors)

4.14 *Milk delivery system.* Cowdenbeath is a branch of Buttercup Dairies, which is a nationwide chain. Staff at Cowdenbeath include the branch manager, Ian Renshaw, two clerks, one storeman and 21 roundsmen (milkmen).

Each round is made up of regular customers who have standing orders for milk and other products. Customers can change their standing orders as they wish – they can order an extra pint or cancel their order for a week while they are away on holiday. In the dairy these variations are known as exception orders. Customers' standing orders, exceptions and account details are recorded in the roundsman's book.

On return to the depot each roundsman gives the storeman a requisition list – his order for the next day. The requisition list is derived from the standing orders, any known exceptions (e.g. holidays) and left-over stock. Goods are booked out to the roundsmen each morning and recorded in the storeman's book.

A customer's weekly bill is calculated from his or her standing order and exception orders for that week. The roundsmen deliver the bills on Thursdays and collect the money on Fridays. When they get back to the depot, the roundsmen count the money and hand it over to one of the clerks, who records it in a cash book. Each roundsman keeps a notebook in which he records the amount booked out to him by the storeman, any stock left over at the end of the day and the amount of money handed over to the clerk.

At the end of each week all the roundsmen have to complete an LCI. This is a form which shows the current financial state of their round. The LCI is given to the clerk, who then checks it against the storeman's book and against the cash book. At the end of the month Ian Renshaw has to send the checked LCIs to Head Office.

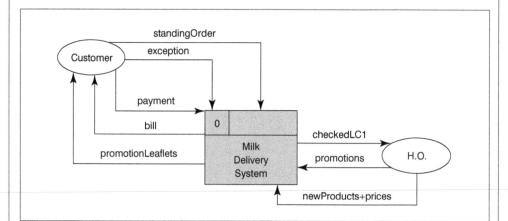

Figure 4.28 Milk delivery system: context CPDFD

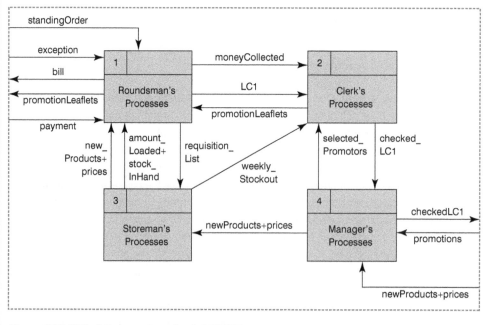

Figure 4.29 Milk delivery system: level 1 CPDFD

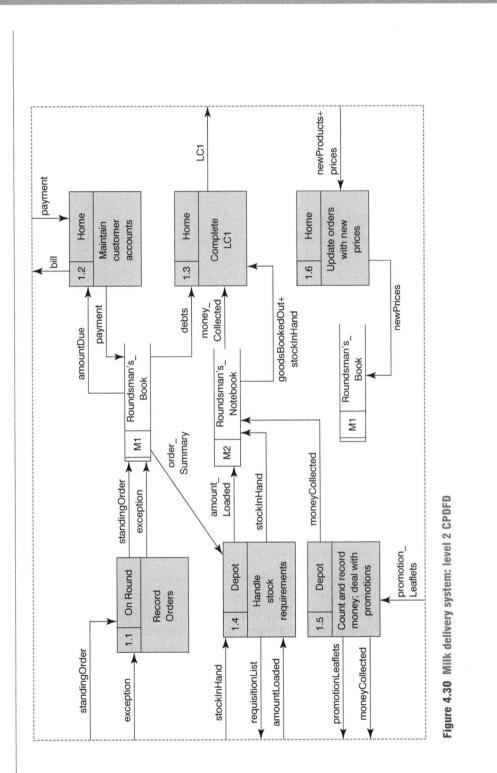

Figure 4.30 Milk delivery system: level 2 CPDFD

From time to time Head Office organizes promotions on certain goods, in order to boost flagging sales. Leaflets about the goods on promotion are distributed to regular customers.

(a) Examine the levelled set of current physical data flow diagrams of the milk delivery system (Figures 4.28–4.30). Note that data flows that cross the system boundary may be represented by arrows that start or end at the system boundary but do not actually cross it.

(b) Draw a level 2 current logical data flow diagram of process 1 of the milk delivery system.

Exercises on process definitions

4.15 In a group, discuss when you would choose to write a process definition using:
- structured English
- decision table
- decision tree.

4.16 Use a decision table to describe the appropriate action for a driver approaching traffic lights in each of the possible combinations of red, amber and green.

4.17 The following is a description of the procedure for determining carpet-fitting charges for carpets bought from Paradise Goods Ltd.

For the purpose of determining delivery charges, customers are divided into two categories: those who are account holders and those who are not. If an account holder buys a carpet costing less than £1000, the fitting charge to be added to the cost of the carpet is £1.25 per square metre. However, if the carpet costs £1000 or more, the fitting charge is £0.75 per square metre. If a customer is not an account holder and the carpet cost is less than £1000, the fitting charge is £1.65 per square metre. For a carpet costing £1000 or more, however, the fitting charge is £1.15 per square metre.

(a) Draw a decision tree showing the process described above.

(b) Describe the same process using structured English.

4.18 Miss Take takes pride in dealing appropriately with all weather conditions. She follows a strict set of rules which ensure that she never lives up to her name. She never goes to work if it is raining in November. If she does go to work she takes an umbrella when it is raining and a coat when it is windy. She always wears a hat when she goes to work, unless it is windy. If it is windy in November she switches on her central heating.

Draw a decision table to show Miss Take's routine.

4.19 (a) Figure E4.3(b) in 'Answers to selected exercises' shows a level 1 data flow diagram describing an automatic ticket machine. Write a process definition in structured English for process 1: 'Calculate ticket price'.

(b) Write a process definition in structured English for process 2: 'Issue ticket'.

References and further reading

Coodland, M. and Slater, C. (1995) *SSADM Version 4: A Practical Approach*, McGraw-Hill, London.

Hawryszkiewycz, I. (2001) *Introduction to Systems Analysis and Design*, 5th edn, Prentice-Hall, a division of Pearson Education, Australia.

Hoffer, J.A., George, J.F. and Valacich, J.S. (2004) *Modern Systems Analysis and Design*, 4th edn, Prentice-Hall, a division of Pearson Education, NJ.

Kendall, E.K. and Kendall, J.E. (2005) *Systems Analysis and Design*, 6th edn, Prentice-Hall, a division of Pearson Education, NJ.

Robinson, B. and Prior, M. (1995) *Systems Analysis Techniques*, International Thomson Computer Press, London. (Unfortunately, not currently in print.)

CHAPTER 5
DATA
DICTIONARY

In this chapter we look at:

- the data dictionary

- notation for a data dictionary

- the Just a Line data dictionary

- content and structure of a data dictionary

After studying this chapter and working through the exercises you will be able to:

- explain the purpose of a data dictionary

- describe a document using data dictionary notation

- use a data dictionary to complement other modelling techniques

- construct and maintain a data dictionary

Introduction

One of the most important tasks in modelling a software system is to identify and provide definitions for all the data in the system. If everyone concerned in the development of the system cannot agree on exactly what they are talking about, then it is very difficult to make any progress. All professional fields have their own specialist vocabulary and within that field meanings are usually well understood. However, in the development of a software system, experts from different fields have to co-operate in tasks such as establishing the problem requirements, designing the solution and developing the user interface. In this situation, it is necessary to agree a formally verified glossary or dictionary to collect and define all the terms being used. In large system development projects today, this glossary is generally constructed and maintained by a **CASE** tool (see Chapter 11).

5.1 Data dictionary

The data dictionary records data about data. It is used to support the other modelling techniques. For example, data dictionary descriptions will be required:

- when capturing all the information given on documents input to the existing system, such as order forms
- when designing output documents, such as reports, mail shots and invoices
- when describing repositories of data in the existing system, such as lists or files, or when designing files for the new system
- when describing the contents of data flows between the processes and data stores on the data flow diagram.

In all of these cases, and many others, a complete detailed description of the data is needed.

A data dictionary can be recorded manually, on paper, or it can be automated, often as part of a **CASE** tool (see Chapter 11). The precise content and layout of the data dictionary varies from one development method to another. The data dictionary technique described here is suitable for describing a small system. A different technique, suitable for larger and more complex systems requiring the cataloguing and cross-referencing of thousands of data items, is described in Kendall and Kendall (2005).

Value of the data dictionary

Whether automated or manual, the data dictionary is an invaluable tool. Data dictionary descriptions are used at the requirements specification, design and implementation stages of system development. The amount of detail required in the descriptions of data changes as the project progresses; normally it is inappropriate to attempt to describe the data with too much low-level detail in the early stages. The data dictionary provides an unambiguous and concise way of recording data about data. It forms a central store of data that supports the information given in other models of the system, not only the **data flow diagrams** but also the **entity-relationship diagram** (see Chapter 6), the **process definitions** (see Chapter 4), the **entity life histories** and **state diagrams** (see Chapter 7). One of the advantages of this is that these models of the system can use short, simple labels to describe data objects. The models are kept uncluttered and readable with no loss of precision, as the labels are cross-referenced to the data dictionary.

Another advantage of using a data dictionary common to all system models is that it encourages consistency between the models; if the same name is used in two models it means the same. Whenever the reader of the model requires more detail about a label, it can be looked up in the data dictionary. For example, the current logical data flow diagram in Figure 4.5 (see Chapter 4) uses simple labels for the data flows **order**, **stockLevel** and **pricedOrder**. The data stores are labelled **Orders** and **ProductInformation**. Each of these labels will have an entry in the data dictionary where a full description will be given. To ensure that all labels are properly documented, the data dictionary should be built up as the data flow diagrams are created.

A well-maintained data dictionary avoids ambiguity about the exact meaning of the words and terms used on a project, and ensures that everyone working on the project is using terminology consistently. The data dictionary also resolves problems of aliases where different people or departments use different names for the same data item. Both names will be entered in the data dictionary and shown to be equivalent.

Notation

There are several notations for recording data dictionaries. The one summarized in Figure 5.1 is simple, but capable of describing the basic configurations in which data items can occur.

A label on a data flow or data store usually names a data group that consists of more than one data element or **_attribute_**. For example, the data flow labelled **customerDetails** in Figure 5.2 describes a data group.

This data group might consist of a title, initials, an address and a telephone number, e.g. Mr A.H. Green, 14 Ferry Road, Littleburgh, Tel. 012 345 6789. This could be described in the data dictionary as:

```
customerDetails = title + {initial} + surname + customerAddress +
                  (customerPhone)
```

Function	Symbol	Description
definition	=	consists of
sequence	+	attributes are joined by +
repitition	{ }	repeated attributes enclosed in braces { }
optionality	( )	optional attributes enclosed in parentheses ( )
selection	[ ]	selection is indicated by enclosing the alternative attributes in square brackets [ ] seperated by a vertical bar, e.g. ["Y" | "N"]
separate	|	separate alternatives in [ ]
value	" "	values given in "..."
comment	*...*	comments between asterisks

Figure 5.1 Notation for a data dictionary

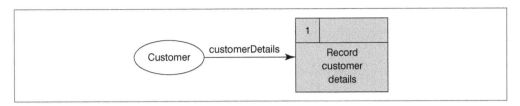

Figure 5.2 Fragment of a data flow diagram to record customer details

The notation must accurately document the way these attributes are combined, i.e. it must express:

- the sequence in which attributes occur – in the above example, **title** then **initials** then **surname**, etc.
- whether an attribute or group of attributes is repeated – the braces round {**initial**} indicate that a customer can have several initials
- whether an attribute is optional – the parentheses round (**customerPhone**) indicate that a customer might or might not have a phone number
- selection from a set of possible values for an attribute – **title** = [**"Mr"** | **"Mrs"** | **"Miss"** | **"Ms"**]; the square brackets indicate a set of possible values for the attribute title; actual values are enclosed within inverted commas and separated by vertical bars
- comments about the entries in the dictionary, e.g. **"local customer"**.

Describing documents

Figure 5.3 shows an order form currently used in the Just a Line system. When a customer phones in an order (or comes into the shop to place an order), the details of the order are recorded, in triplicate, on this form.

This order form is represented in several places on the current physical data flow diagram (Figure 4.17 in Chapter 4); it has different labels according to what stage it is at and how much information has been recorded on it. Initially, it corresponds to the flow labelled **order**, and is used to record the details of the customer's order before it is priced. In the data dictionary, assuming details of pricing are added at a later stage, the details of the order might be described by the entry:

```
order = (title) + {initial} + surname + customerAddress +
        (customerPhone) + (deliveryAddress) + dateRequired +
        {productDescription + packetSize + numberOfPackets}
```

This description is rather long, difficult to read and difficult to reproduce without introducing transcription errors. An order can be described more elegantly by splitting the single description into a number of shorter ones:

```
order =            orderHeader + {orderLine}
orderHeader =      customerDetails + (deliveryAddress) +
                   dateRequired
customerDetails =  (title) + {initial} + surname +
                   customerAddress + (customerPhone)
orderLine =        productDescription + packetSize +
                   numberOfPackets
```

Just a Line ...

ORDER FORM

Mr/Mrs/Miss/Ms _____

Address _____

Delivery Address _____

Date Required _____ Telephone _____

Description	Packet Size	No. Packets	Price	£	p
			Total Cost		
			Delivery Charge		
			Amount Owing		

Figure 5.3 Just a Line order form

This description means that an **order** consists of an **orderHeader** and one or more **orderLines**. Notice that in one respect the data dictionary can be used like a normal English dictionary; terms that are used in the description of one entry may themselves be entries in the dictionary. Thus, **orderHeader** is used to describe **order** and can also be looked up in the *data dictionary*; orderHeader is an example of a data structure – a group of data attributes referred to by a label. Notice how this data structure is used again in the description of **pricedOrder** (below). Data structures are used for convenience in the data dictionary, allowing a long string of attributes to be replaced with a label that describes them. This technique keeps descriptions concise and more readable, avoids repetition of long strings of attributes and helps to prevent transcription errors creeping in.

Once the prices have been added to the order form it becomes a **pricedOrder**. The corresponding entry in the data dictionary reads:

```
pricedOrder = orderHeader + {pricedOrderLine} + totalCost +
               (deliveryCharge *if total cost < £30*) + amountOwing
pricedOrderLine = orderLine + costPerPacket + costOfLine
```

Notice the following points.

■ The economical use of the already defined data structures **orderHeader** and **orderLine**.
■ All the essential information recorded on the order form currently used in the Just a Line system has been captured in this data dictionary description.
■ The notational convention used here is the same as that used for labelling data flow diagrams – labels for data flows, elements and structures are written in lower case. Where a label consists of more than one word, the words are run together with no spaces but with a capital letter at the start of the second and each subsequent word. Names of data stores are written as for data flows except that the first word is also capitalized.

What is an appropriate level of detail?

The amount of detail appropriate for a data dictionary depends on how the data dictionary will be used. A data dictionary for a current system, for example, will normally go into less detail than a data dictionary built at the detailed design stage. If the data dictionary is to be used simply to document the system developer's understanding of current processing and to support discussions with the client, a label like **dateOfOrder** will be self-explanatory; no more detail is required. At the detailed design stage, however, the system developer is thinking ahead to the detail required by a programmer. Decisions have to be made that were unnecessary earlier: decisions, for example, about input and output formats for dates. Will the system, for instance, use '20/10/2005' as input format and '20th October 2005' as output format, i.e. decisions must be made about the format or 'picture' of each data element and about its permissible range of values. This information will be required when input documents or screens are being designed and when input validation checks are specified, and will be recorded in the data dictionary.

Just a Line: current logical data dictionary

A data dictionary to support the current logical data flow diagrams, shown in Figures 4.4, 4.5 and 4.6, of the Just a Line system is listed below:

```
checkedSupplierInvoice = supplierName + supplierAddress +
                         {pricedStockOrderLine} + totalCostOfGoods +
                         (supplierDeliveryCharge) + totalAmountOwing
                         *checked against stock delivered*
confirmationOfDelivery = supplierInvoice + JustaLineSignature
customerDetails = title + {[forename | initial]} + surname +
                  customerAddress + (customerPhone)
customerPayment = pricedOrder + customerSignature + payment
deliveryNote = pricedOrder
invoice = pricedOrder
newStockLevel = productDescription + quantityDelivered
order = orderHeader + {orderLine}
orderHeader = customerDetails + (deliveryAddress) + dateRequired
orderLine = productDescription + packetSize + numberOfPackets
Orders = *data store* {pricedOrder}
```

```
price = productDescription + packetSize + costPerPacket *in the
            current system price must contain the same information as
            productDetails*
pricedOrder = orderHeader + {pricedOrderLine} + totalCost +
                (deliveryCharge *if total cost < £30*) + amountOwing
pricedOrderLine = orderLine + costPerPacket + costOfLine
pricedStockOrderLine = stockOrderLine + costPerUnit + costOfLine
productDetails = productDescription + packetSize + costPerPacket
ProductInformation = *data store* {productInformation} *i.e.
                        currently the design list, the price list and
                        the contents of the store*
ProductInformation = supplierName + supplierAddress + {productCode +
                        productDescription + packetSize + costPerPacket
                        + quantityOnHand}
quantityRequired = *quantity of product required to fill outstanding
                    orders*
stockLevel = productDescription + quantityOnHand *number of packets
                left in the store*
stockOrder = supplierName + supplierAddress + {stockOrderLine}
stockOrderLine = {productCode + productDescription + packetSize +
                    numberOfPackets}
stockOrderRequirements = supplierName + supplierAddress
                            + {productCode + packetSize + stockLevel}
supplierInvoice = supplierName + supplierAddress
                    + {pricedStockOrderLine} + totalCostOfGoods
                    + (supplierDeliveryCharge) + totalAmountOwing
supplierPayment = supplierInvoice + payment
```

This data dictionary is short and uncomplicated. It is therefore sufficiently structured if kept simply in alphabetic order. Notice that **order**, **pricedOrder**, **invoice** and **deliveryNote** refer to the same piece of paper at different stages in the system.

Contents of data dictionary

A simple data dictionary can be kept as an alphabetic list, as above. A more sophisticated data dictionary may have separate sections for data elements, data structures, data flows, data stores and entity descriptions (see Chapter 6).

Data elements

Data elements are the primitives of the data dictionary. Examples in the description of the Just a Line order form are **dateRequired**, **costPerPacket** and **productDescription**. These are the smallest data items used in the system and are not further decomposed. A full data dictionary description of a data element in the later stages of system development will include a description, information about the format (or picture) of the data element, the range of permissible values, aliases and an indication of where the data element is used.

Data structures, data stores and data flows

A data structure is a named (labelled) list. A data structure is a combination of one or more data elements, or a combination of data elements and data structures – all of which are defined elsewhere in the data dictionary. Data flows and data stores are data structures. A data structure may correspond to a data flow or a data store on a data flow diagram, but it does not have to. It allows us simply and unambiguously to describe a sequence of data elements. In a full data dictionary each data structure will be described in terms of its content with additional cross-referenced information about which diagrams use it, which documents it describes, where its elements are described and where it is incorporated into other data structures.

A full data dictionary for even a modest-sized system can run to hundreds of pages. A more manageable approach, for anything other than a very small system, is to use a CASE tool whose repository sets up and manages all aspects of the data dictionary. This is especially useful for ensuring that any changes that affect the data dictionary filter through to all the models concerned (see Chapter 11).

Summary

A data dictionary provides a central store of data about data. It allows the system developer to describe data flows and stores by simple names, keeping the data flow diagrams readable and uncluttered. A data dictionary solves many communication problems – everyone working on the same project knows the exact meaning of the words and terms used. The data dictionary is also used to record data about the data model, the process definitions, the entity life histories and the state diagrams.

Exercises and topics for discussion
What can you remember?

You will find all the answers in the chapter

a) What is the purpose of a data dictionary?
b) In what ways does a data dictionary support other modelling techniques?
c) How does the data dictionary show sequence, repetition, optionality and selection?
d) How does the data dictionary indicate values and comments?
e) Define the terms data element and data structure. Give examples of both.
f) Why are data structures used in a data dictionary?

5.1 In the Just a Line case study, check that all the data flows and data stores on the current logical data flow diagrams are defined in the data dictionary.
5.2 Using the information given in the interview with Sue and Harry, and information recorded in the current logical data dictionary of the Just a Line system, write data dictionary descriptions of the data stores:

```
Store
DesignList
PriceList
```

shown on the current physical data flow diagram in Figure 4.17.

5.3 Write a data dictionary to support the data flow diagram of the automatic ticket machine system case study, Exercise 4.3.

5.4 The document in Figure 5.4 is an example of a customer's monthly statement from the department store credit system described in Exercise 4.5. Describe this document in data dictionary notation.

Joshua Prichards

Joshua Prichard & Sons Ltd
Customers' Accounts
PO Box 12, Cambridge, England CB1 1PT
Telephone (01223) 654321

Statement of Account

Statement No.	Card No.	Statement Date
1737	B37787	31/12/05

Mrs V. I. Paterson
8 Walden Road
Saffron Walden
Essex

Date	Department/Reference		Debit	Credit
3/12/05	L /124814	A	£23.95	
8/12/05	D /143712	A	£2145	
15/12/05	A /141497	A	£7.99	
18/12/05	TC2189	A		£7.99

Payment Due	Credit Balance
£45.40	

Figure 5.4 Monthly customer statement

5.5 Using the case study notes and current physical DFDs of the milk delivery system specified in Exercise 4.14 in Chapter 4, write data dictionary descriptions for the flows **standingOrder**, **exception** and for the data store **Roundsman'sBook**.

References and further reading

Kendall, E.K. and Kendall, J.E. (2005) *Systems Analysis and Design*, 6th edn, Prentice-Hall, a division of Pearson Education, NJ.

CHAPTER 6
DATA
MODELLING

In this chapter we look at:

- the principles of data modelling

- the entity-relationship model

- normalization

After studying this chapter and working through the exercises you will be able to:

- explain what data modelling aims to achieve

- draw an entity-relationship model for a simple system

- normalize data up to third normal form

Introduction

A *data model* is often developed in parallel with the data flow diagram and given equal emphasis. Some methodologies do the data flow diagram before the data model, some do the data model first. Our own view is that the order is not significant; both are tools the developer should use where and when they will be helpful. The data flow diagram is concerned with the question, *What does the system do with the data?* The data model is concerned with the question, *What data does the system need to store and what is the most efficient way of organizing it?* Stored data requirements are not ignored by the data flow diagram: stored data is modelled in the data stores. However, when designing data stores, the system developer is not attempting to find an efficient way of organizing the data, but is merely identifying the stored data items and grouping them in some appropriate way. In a data flow diagram of the *current physical system* the data stores correspond to identifiable physical repositories of data, e.g. ledger books, filing cabinets and reference manuals. A data flow diagram of the **current logical system** abstracts the data from the current implementation and groups data that seems to be associated in some way, each group being represented by a data store.

Data modelling evolved as part of relational database design. In the mid-20th century commercial companies usually organized their data into separate mainfiles for each application system. An insurance company, for example, might have a car insurance mainfile, a house insurance mainfile, a life assurance mainfile and a pension mainfile. Details about one customer were often stored in more than one mainfile. This meant that if a customer, Miss Philips, changed her address when she returned the renewal notice for her car insurance, her record on the car insurance mainfile would be updated, but if she had records on any of the other mainfile, they would probably not be updated. The complications of trying to ensure that duplicated data was maintained consistently was what led to the development of a centralized system of storing data in a company database. As far as was possible, data was stored in one place only. Data was grouped logically rather than by application, e.g. all the data about customers together, all the data about products together, and so on. Rules were drawn up to ensure that the data was correctly entered and maintained. These rules were enforced by the database software. The database was then used by all of the application systems.

Data modelling uses two separate techniques to achieve a satisfactory set of entities: **entity-relationship (E-R) modelling** and **normalization**. We discuss data modelling in three sections. The first discusses issues common to E-R modelling and normalization, the second discusses E-R modelling, and the third discusses normalization.

6.1 Principles of data modelling

The data model sets out to capture all the data that the system needs to store and to organize the data into an efficient structure. It goes about this by pursuing four main objectives.

1. The identification of the data objects or entities in the system, their structure, and the **relationships** between entities.
2. The construction of a model of the stored data requirements of the system that is independent of specific processing requirements.

3. The construction of a robust data model, i.e. of a minimal model of the data required to be stored by the system.

4. The construction of a logical model of the data, i.e. a model that is not concerned with how the data storage will be, or is currently, physically implemented.

1. Identification of entities

Data modelling aims to identify the system entities, i.e. items about which the system needs to store data (e.g. customers, orders, products, etc.). Data modelling also shows the internal structure of these entities, i.e. describes the details stored about each entity (e.g. for each customer store the customerName, customerAddress, customerPhone). The system developer also uses the data model to show the relationships between entities, i.e. show any relationships between entities that are significant to the system. For example, customers place orders and therefore there is a significant relationship between customers and orders.

2. Process-independent model

Whereas the data flow diagram concentrates on what happens to the system data, the data model concentrates on the properties of the data itself. During data modelling we aim to collect all the data that needs to be stored for the system to operate and to organize that data into a sound structure that will perform well given the current processing requirements and that will efficiently accommodate future processing requirements. We concentrate on the data itself and organize it according to its own internal logic rather than what the system currently wants to do with it. This is because studies have shown that the stored data in a system is more stable than the processing requirements of a system. During the lifetime of a system the processing requirements will almost inevitably evolve as the system changes to meet a changing environment. For example, the system may be required to produce extra reports, include a new set of functions (perhaps the accounting functions) or do its calculations differently. However, the data does not tend to change significantly over the system life span. A correct data model provides a solid foundation on which to build the rest of the system.

3. Minimal model

Data modelling was originally devised for *database* design. Database design stresses the importance of a process-independent approach to the organization of the stored data and the importance of allowing as little *redundant* data as is feasible – as far as is possible one item of data should be stored in one and only one place. If the same item of data is stored in more than one place, it is unnecessarily using up storage space and may result in the system having discrepancies between data that has been updated in one place but not in another. Another potential source of discrepancy is stored data that has been derived (i.e. calculated) from other data stored in the system. A robust data model is one that does not permit this sort of discrepancy.

4. Logical model

Data modelling is not concerned with how the data is physically stored, either in the current system or in the computer-based system being developed. The reason for this is that the most efficient way of organizing data on, for example, a card index in a manual system, will almost certainly be very different from the most efficient way of storing data in a computer-based system. The data model aims to develop an efficient model of the data independently of how it is implemented.

Two techniques: E-R modelling and normalization

Data modelling uses two quite distinct techniques to achieve a satisfactory organization of the system's data: a top-down approach, which we refer to as E-R modelling, and a bottom-up approach, which we refer to as normalization. Each technique produces a data model, but the models are underpinned by different degrees of rigour. Both aim to achieve the goals listed above. E-R modelling begins by looking for the data groups in the system. It starts with the question, *Who or what do we need to store data about?* It provides a useful first attempt to organize the data but results in a data model that may be incomplete and may not have achieved the most efficient organization of the data. Specification of the relationships between entities is based on common sense rather than mathematical rigour. However, this first data model can provide a sensible starting point. It is also useful as a check against the normalized model. The second technique, normalization, starts bottom-up, by looking at the smallest individual items of data recorded by the system. This technique takes as its starting point the mass of data items collected during the fact-finding stages of analysis: all system data can be subjected to the normalization process – inputs, outputs, files, lists, reports. Following a series of well-defined systematic rules, normalization will structure these data items into a related set of entities, forming a rigorously constructed data model. The data model achieved using normalization provides a structure ready for implementation.

Terminology and notation

Data modelling has its own terminology: it describes data in terms of entities, occurrences of entities, attributes of entities, values of entities, entity keys and relationships between entities.

Entities

An *entity* is an object about which the system needs to store data, such as Customer or Product. It is represented as a box on the E-R model diagram with the name of the entity inside, as in Figure 6.1.

An *occurrence of an entity* itemizes the values relating to one *instance* of an entity – for example, one particular customer or product. The entity itself represents a number of entity occurrences, all the customers or all the products in the system; it is sometimes referred to as an *entity type*. In this book an entity or entity type is always referred to with an initial capital letter (e.g. Customer) and an occurrence of that entity is referred to in lower case (e.g. customer).

The *attributes of an entity* are the data items or elements that make up that entity. For example, customerName, customerAddress and creditStatus might be three *attributes* of the entity Customer. Entities often represent something in the real world, as is the case with the entity Customer. However, the attributes we record about a customer are only those of

Figure 6.1 Representation of entities

significance to the system. The Just a Line system will not record, for example, a customer's taste in music or literature. These attributes might be of significance in a system processing orders for records or books, but not in one that deals with mail-order card sales.

The *value of an attribute* is the value of that attribute for a particular entity occurrence; for example, John Barrett might be the value of the attribute customerName.

To summarize: an entity such as Customer represents a group of customers and lists all the attributes that we want to collect about our customers such as customerName and customerAddress – it names this group of attributes.

An easy way to distinguish between an entity and an occurrence of an entity is to think of what we do when we draw up a list – for example, a list of the players in a university football team. On a sheet of paper we write a heading, e.g. Team – this is the entity name. Then we divide the page into columns with a heading at the top of each column, e.g. name, address, telephone number, mobile number – these are the entity attributes. When we fill in the details about each player, we are recording values for each of the entity attributes. The details about one player will take up one row; each row, therefore, represents an occurrence of the entity Team.

Primary keys

The **primary key** of an entity is an attribute or set of attributes whose values uniquely identify one occurrence of that entity, e.g. one customer or one product. If the system is to work, each occurrence of an entity (or each record of a computer-based implementation) must be distinguishable from the others. It must be possible to identify it and to extract the relevant stored details. To do this we need a distinguishing feature for that occurrence. Normally, this will be one of its attributes, but sometimes a combination of more than one attribute has to be used. For example, in an entity Student, a collection of students may be distinguishable from each other by name. However, if their name does not uniquely identify each student, then a combination of name and dateOfBirth may work. Failing this, we could give each student a unique studentNumber. We would then be able to guarantee that, given the appropriate studentNumber, we could extract the rest of the stored details about that student – the value of the studentNumber **determines** the value of the other data items in the entity. All keys that uniquely identify an occurrence of an entity are known as **candidate keys**. If a combination of name and dateOfBirth can be relied upon to uniquely identify each student, then this combination of attributes is one candidate key; studentNumber is another candidate key. If studentNumber is what we decide to use as the key for this entity, it becomes the primary key. In the Just a Line system; customer#[1] might be used as the primary key to Customer if we design the customer numbers in such a way that no two customers have the same customer#. Entity primary keys are underlined as shown here:

```
Customer (customer#, initial, surname, customerAddress,
customerPhone)
```

Entities or entity types are conventionally described as above: the entity name followed by the attributes enclosed in parentheses: ( ). Note that this differs from data dictionary notation (see Chapter 5) for which parentheses mean that attribute values are optional.

---

[1] The symbol # means number.

To fulfil its function of uniquely identifying each instance of an entity, a primary key should have the following properties.

- A primary key must be unique for each instance (or occurrence) of an entity.
- A primary key must always have a definite value – it must never be allowed to have a null value, otherwise the possibility arises of several entity occurrences having the same primary key. As soon as an instance of an entity is created it must have a value assigned for its primary key, e.g. a new student must immediately be given a student number.
- A primary key should not contain an attribute that is liable to change; e.g. student name is not an ideal primary key, even when combined with date of birth, as it may change – for example, on marriage.
- System developers should maintain control over the primary keys. One system team, which was developing a milk delivery system, had serious problems because it was persuaded to use customers' telephone numbers as the primary key. The team did not realize that the telephone company might change the numbers, or that there would be households such as student flats where there might be several customers with the same telephone number.

Foreign keys

An entity may contain as one of its attributes a data item that is the primary key of another entity. For example, Figure 6.2 shows a list of occurrences of an entity Employee while Figure 6.3 shows a list of occurrences of an entity Department. One of the attributes of the Employee entity is deptNo, which is the primary key of the entity Department; this makes deptNo a **foreign key** in the Employee entity. Foreign keys act as links or navigation routes between related entities.

Relationships

A relationship is a link between two entities, which is significant for the system. For example, a customer places an order – 'places' describes a relationship between a Customer and an Order. This is a relationship in the real world and one that would be significant in an order-

EmpNo	Name	StartDate	Scale	DeptNo	ExtNo	OfficeNo
143	M. Cobby	12/04/86	S1	42	2345	663
267	S. Bedbrook	01/06/82	GS2	42	4271	663
281	B. Watts	23/10/91	P3	21	9478	251
296	D. Hawkins	01/09/92	P2	16	4364	382
341	I. Oxford	07/10/89	GS2	33		337
367	A. Varty	10/09/70	D4	33	3216	338

Figure 6.2 Employee entity

DeptNo	DeptName	HeadOfDept
16	English	A. C. Taylor
21	French	C. F. A. Andrews
33	Computer Science	M. H. Martin
42	Maths	J. F. Smith

Figure 6.3 Department entity

processing system. On an E-R diagram the relationship is modelled by a line between the two entities concerned, as in Figure 6.4.

It is important to capture all the significant relationships when designing information systems because they trace the access from one entity occurrence to another. If a relationship exists between Customer and Order, this implies that for each occurrence of Order we can trace the customer who placed that order; similarly, for each occurrence of Customer (i.e. each customer in the system), we can trace all the associated occurrences of Order (i.e. all the orders they placed). Entities must be linked to reflect all significant real-life relationships so that when implemented on a computer-based system, all processing requirements, both now and in the future, can be satisfied. Sometimes it is useful to name the relationship between entities as in Figure 6.5.

The degree of a relationship

The relationship between Customer and Order is one to many, in that any one customer may place many (one or more) orders and any one order is placed by just one customer. A relationship is indicated by a line between entities on the E-R diagram. The **degree of the relationship** is indicated by a **crow's foot** on the many end of the line. Relationships between entities can be:

- one to one
- one to many
- many to many.

The terms **multiplicity** and **cardinality** are sometimes used to refer to the degree of relationships.

Figure 6.6 gives an example of each type of relationship using E-R diagrams. Figure 6.7 shows the same relationships in terms of entity occurrences; this type of diagram is known as an occurrence diagram (see below).

One to one: Assuming that a husband has one wife (at a time) and a wife has one husband (at a time), we have the situation modelled by the E-R diagram in Figure 6.6. More formally, any occurrence of the entity Husband (i.e. any husband) is associated with only one occurrence of the entity Wife (i.e. one wife). Figure 6.7 illustrates one-to-one relationships between occurrences of the entities Husband and Wife.

Figure 6.4 Representation of relationship

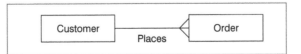

Figure 6.5 Named relationship

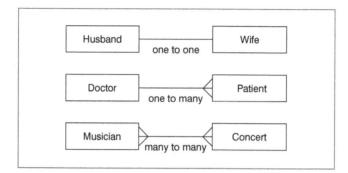

Figure 6.6 Degree of relationships: E-R diagrams

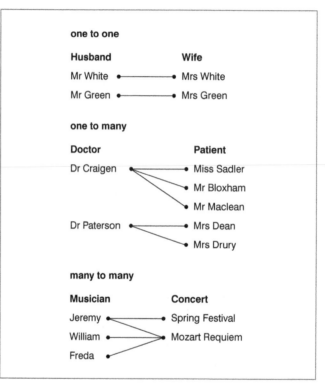

Figure 6.7 Degree of relationships: occurrence diagrams

One to many: A doctor has many patients but the rules of the practice dictate that a patient is registered with only one doctor.

Many to many: A musician plays in many concerts and a concert is performed by many musicians.

Optional and mandatory relationships: A relationship between two entities can be mandatory or optional. A relationship is optional if, where a relationship exists between two entities, it is possible for an occurrence of one entity to exist without being associated with an occurrence of the other entity. The relationship between the entity Mother and the entity Son is mandatory for

Son and optional for Mother – a son cannot exist without having (at some stage) a mother but a mother does not have to have a son, since all her children may be daughters. An optional relationship is shown by a dashed line and a mandatory relationship by a solid line, as in Figure 6.8. Where a relationship is optional for one entity and mandatory for the other, the dashed line goes next to the entity that does not have to participate in the relationship, in this case Mother. A relationship between two entities is mandatory if every occurrence of one entity must participate in a relationship with at least one occurrence of the other entity. In Figure 6.9 a mandatory relationship exists between Student and Course – every student must take at least one course and every course must be studied by at least one student.

The notation used here for drawing E-R models is one of several different notations that serve the same purpose. However, it is only the symbols in the various notations that are different, the underlying logic is the same. A summary of the notation used in this book is given in Figure 6.10.

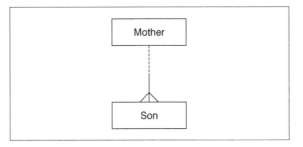

Figure 6.8 Optional relationship for Mother, mandatory relationship for Son

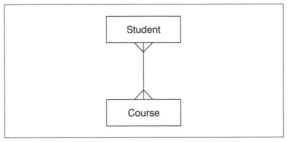

Figure 6.9 Mandatory relationship for both Student and Course

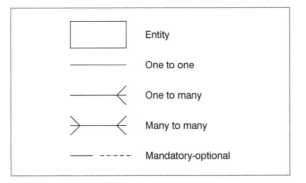

Figure 6.10 Summary of notation for data modelling

Occurrence diagrams can be useful for working out or explaining the degree of relationships. In an occurrence diagram each occurrence of an entity is represented by a named dot. Relationships are represented by lines drawn between the dots. For example, in Figure 6.7 we use dots to represent the patients Miss Sadler, Mr Bloxham, Mr Maclean, Mrs Dean and Mrs Drury and the doctors Dr Craigen and Dr Paterson. The lines drawn between the dots tell us which patient is registered with which doctor. The diagram shows us clearly that each doctor has several patients, but that each patient is registered only with one doctor. Occurrence diagrams are clear, but it would be unrealistic to try to plot out all data relationships using this technique – it is too laborious. Once the nature of the relationships between occurrences of an entity has been understood, they can be represented more concisely by an E-R diagram. The information about patients and doctors shown in Figure 6.7 is summarized in the one-to-many E-R diagram in Figure 6.6.

Enterprise rules

E-R modelling and normalization give us guidelines and rules to help us construct a satisfactory data model. However, if the model is to be correct, we need a sound understanding of what data the system requires and how the data items relate to each other: the rules that govern the data in the system we are modelling. These rules are sometimes referred to as **enterprise rules**. Enterprise rules for the library system described below include such information as:

■ a library member is uniquely identified by his or her member number
■ a book is identified by its ISBN (International Standard Book Number)
■ the library may have several copies of one book; individual copies are identified by a library ID number.

The system developer must have a thorough understanding of these rules in order to develop a successful data model.

6.2 Entity-relationship modelling
Practical guidelines

The principles of data modelling, listed near the beginning of this chapter, are common to both the techniques of E-R modelling and normalization. In practical terms we achieve the purpose of data modelling by building a data model with various well-defined characteristics, as outlined below.

1. The first is that data items should be put into logical groups, groups of data items that 'go together'. For example, we might group together all the data relating to Customer – name, address, etc. – just as in a manual filing system we might have a card file of customers, each card having the details about one customer. Failure to group data items into groups that are mutually dependent can cause unexpected problems. For example, in the current system at Just a Line there is no customer data held except on the orders. It follows that, as far as the system is concerned, a customer ceases to exist for any period that he or she does not have an order being processed. This does not matter much in the existing system, but the new system is required to send out mail shots to all customers. Without a permanent file of customers' details this is impossible.

2. The second characteristic is that for each data group, or entity, there should be a primary key that uniquely identifies individual members of that entity. For example, a customer's name might uniquely identify an individual customer in a small system. In a larger system with more customers we might have to introduce a customer number to uniquely identify each customer.

3. The third characteristic of a good data model is that there should be no redundant data in the model. Data is deemed to be redundant in the following situations.

   (a) The data is never used by the system.

   (b) The same data items (e.g. customerName, customerAddress) are stored in more than one place in the system.

   (c) Data in one place can be derived from data held in another place in the system (e.g. the total order price can be derived from the individual item prices and quantities ordered).

The first of these situations just wastes space. The other two can cause real problems. If a customer's address is stored in more than one place in the system then, if the customer changes address, we have to track down and update all copies of the address. If we miss some, we will have data in the system that is out of date.

If the system stores **derived data**, this both wastes space and can create situations where the system is storing data that is out of date and therefore incorrect. For example, a system might have customers with regular standing orders. If we decide to store the total order price with these orders we are storing derived data. The total order price can be derived from the quantity and the individual item prices. The danger is that if the product prices change the total order price will be out of date.

To summarize, when designing a data model we must:

- put data items into logical groups that go together
- ensure that there is enough information to identify a particular instance of a data group, e.g. can we uniquely identify one particular customer record?
- ensure that each data item is stored only once
- ensure that each data item needs to be stored, i.e. cannot be derived.

E-R modelling starts by identifying the objects about which the system needs to store data. It then investigates the attributes of each entity and the relationships between the entities. These objects can then be refined using the rules outlined above.

How to build an E-R model

The steps in building an E-R model are outlined below and then illustrated with an example from a simplified library case study.

1. To build the E-R model, look first at the system and identify data objects (entities) about which the system needs to store data. The problem definition (see Chapter 2), or any other written summary of the system requirements, will be helpful here.

2. For each entity, list its attributes.
3. Put the data into logical groups.
4. Identify primary keys.
5. Eliminate redundant data.
6. Investigate and record relationships. We need to be able to link entities so that all significant real-life relationships are captured.
7. Check the entity descriptions against the data dictionary descriptions of data stores, inputs and outputs, data flows, etc. Make sure that each process on the data flow diagram has available to it all the data it needs, i.e. make sure that all the data the system needs to function has been captured.

We will work through the process of building an E-R model using a library system as an example. We will then build an E-R model of the Just a Line system.

The library system

A library keeps records of loans, books and members. It stores members' names, addresses, status (junior or senior), loan limit (number of books a member may borrow) and date of birth; members are given individual member numbers when they join the library. The library also stores information about its books: title, authors, publisher, publication date, ISBN and purchase price. As some books are very popular, the library often buys several copies of the same book. All loans are for three weeks. The library needs to be able to record, edit and delete member details; record, edit and delete book details; record loans and returns; and reserve books. Overdue notices are to be sent when books are overdue. It also wants its library system to automatically update member status. The system must also record the current price of a book.

The following enterprise rules can be applied to the library data.

- Individual copies of books are identified by library ID number; when recording loans, the library needs to be able to identify which copy of the book has been borrowed. The book itself, i.e. a specific title/author combination, is identified by its ISBN.
- Members are identified by member number.
- Member status determines how many books a member may borrow – the 'loan limit'. To keep the example simple, we are ignoring issues relating to fines for overdue books.

Step 1. Identify entities

Using the steps identified above, the first thing to do is to pick out the entities in the library system; the things we need to store data about. We know we have to store data about members, books, loans and reservations. Our initial list of entities, therefore, would be:

```
Member
Book
Loan
Reservation
```

Step 2. List attributes

The description of the library system gives us quite a lot of clues about what attributes we need to store for each of our entities. We are told that the system must store the following attributes for members and books:

```
Member (member#, memberName, address, status, loanLimit,
dateOfBirth)
Book (title, authors, publisher, publicationDate, libraryID#, ISBN,
purchasePrice, currentPrice)
```

The system must record loans. The library wants to know which copy has been borrowed; the enterprise rules tell us that this can be identified by the library ID number. The member can be identified by member number. We need to know when the book is due back so that we can send overdue notices:

```
Loan (member#, libraryID#, dueDate)
```

For a reservation, the member does not care which copy of the book he reserves – he wants the first one that comes back. We would therefore record the ISBN, as this relates to all copies of the book. We will also store a reservation date to tell us how long to keep trying to reserve the book.

```
Reservation (member#, ISBN, reservationDate)
```

Step 3. Put the data into logical groups

Member, Loan and Reservation already form logical groups of data. Book seems to contain some data that relates to copies of books. The system needs to identify and store data about copies of books, but for each copy we don't need to store all of the book information. If we have five copies of *The Hitchhiker's Guide to the Galaxy* by Douglas Adams (Pan, 1979), we don't want to store data in the format shown in Figure 6.11 as this would mean we were storing redundant data.

For individual copies of books we need only store the library ID number and a reference to the Book entity where the rest of the details about the book are stored; we can do the latter by using the ISBN. The other attribute that relates to the copy rather than the book is the price the

Library ID#	ISBN	Title	Author	Publisher	Pubn Date	...
123	0330258648	Hitchhiker's Guide to the Galaxy	D. Adams	Pan	1979	
234	0330258648	Hitchhiker's Guide to the Galaxy	D. Adams	Pan	1979	
345	0330258648	Hitchhiker's Guide to the Galaxy	D. Adams	Pan	1979	
456	0330258648	Hitchhiker's Guide to the Galaxy	D. Adams	Pan	1979	
567	0330258648	Hitchhiker's Guide to the Galaxy	D. Adams	Pan	1979	

Figure 6.11 Data storage illustrating redundancy

library paid for that copy – the purchase price; the current price should be stored with the book data. This splits the Book entity into two entities, Book and Copy, and the book data into two logical groups, book data and copy data:

```
Book (ISBN, title, authors, publisher, publicationDate,
currentPrice)
Copy (libraryID#, ISBN, purchasePrice)
```

Step 4. Identify primary keys

Primary keys are dictated to a large extent by the enterprise rules for the library. We know that members are identified by member number, books by ISBN, copies by library ID number. If we assume that loan records are deleted from the system as soon as a book is returned, then library ID number forms a unique identifier for a loan record, with member# as a foreign key, so that we can trace the borrower. However, if historical records of loans are kept, we could have more than one loan record with the same member number and library ID number, as the same member might borrow the same copy on more than one occasion. In this case the due date will have to be included as part of the primary key. The same condition applies to reservation records: if reservation records are deleted from the system as soon as they have served their purpose – the reserved book has been collected or is no longer required – then the combination of the attributes member number and ISBN forms a satisfactory key. However, if historical records are kept, we need to include reservationDate as part of the key – members may reserve the same book more than once. In this case we have decided to keep historical records as the library wants statistics about the most popular books:

```
Member (member#, memberName, address, status, loanLimit,
dateOfBirth)
Book (ISBN, title, authors, publisher, publicationDate,
currentPrice)
Copy (libraryID#, ISBN, purchasePrice)
Loan (libraryID#, dueDate, member#)
Reservation (member#, ISBN, reservationDate)
```

Step 5. Eliminate redundant data

We have already done some elimination of redundant data by separating books and copies in step 3. In that case it was repeating data that was eliminated – by separating Book and Copy we avoided storing the details of title, authors, publisher, etc., for every copy of the book. However, we could still have a problem with repeating data in the Book entity if a book has more than one author. Consider the book *Object-Oriented Modeling and Design* by J. Rumbaugh, M. Blaha, W. Premerlani, F. Eddy and W. Lorensen – five authors. If we were to write this information on a piece of paper, we might arrange it as in Figure 6.12.

However, this is not a format that can be sensibly or easily used by a computer system; it is not recommended practice to have one attribute that repeats relative to the others. If we want to keep the author data in the Book entity, we would have to allow enough space for five or even more authors, but this wastes space when there is only one author. Alternatively we could allow for only one author per record and repeat the ISBN and title for each author, as in Figure 6.13.

ISBN	Title	Author	...
0136300545	Object-Oriented Modeling and Design	J.Rumbaugh M.Blaha W.Premerlani F.Eddy W.Lorensen	

Figure 6.12 Paper representation of book title and authors

ISBN	Title	Author	...
0136300545	Object-Oriented Modeling and Design	J.Rumbaugh	
0136300545	Object-Oriented Modeling and Design	M.Blaha	
0136300545	Object-Oriented Modeling and Design	W.Premerlani	
0136300545	Object-Oriented Modeling and Design	F.Eddy	
0136300545	Object-Oriented Modeling and Design	W.Lorensen	

Figure 6.13 Problematic data storage

However, this produces repeated data and means that the ISBN no longer identifies individual records. The solution is to put the data about the authors into a separate entity with a reference to the Book entity, so that we do not lose the link between the authors and the book they wrote:

```
Author (ISBN, author)
```

For the above example there would be five Author entity occurrences with the same ISBN. As an author may write several books, as well as co-authoring the same book, the key would have to be ISBN plus author. The Book entity would no longer contain any author information:

```
Book (ISBN, title, publisher, publicationDate, currentPrice)
Author (ISBN, author)
```

We also have a problem with redundant data in the Member entity:

```
Member (member#, memberName, address, status, loanLimit,
dateOfBirth)
```

If we store both status and date of birth, we are storing derivable data because status can be derived from date of birth. Members under 16 are classified as juniors; members who are 16 or over are seniors. Loan limit is derivable from status – four books for junior, eight for senior. We only need store date of birth. This gives us the final set of entities:

```
Member (member#, memberName, address, dateOfBirth)
Book (ISBN, title, publisher, publicationDate, currentPrice)
```

```
Author (ISBN, author)
Copy (libraryID#, ISBN, purchasePrice)
Loan (libraryID#, dueDate, member#)
Reservation (member#, ISBN, reservation Date)
```

Step 6. Investigate and record relationships

The enterprise rules tell us that:

- a member can make many reservations, but a reservation is only for one member
- a member can take out several loans, but a loan is only for one member
- a loan is for one copy of a book, but a copy may be on many loans (over time)
- a book may be reserved many times, but a reservation is only for one book
- a book can have several copies, but a copy is only associated with one book
- a book can have several authors, but an occurrence of an author/ISBN combination is only associated with one book.

This links the entities as shown in Figure 6.14.

Requirements list for Just a Line

We will now construct an E-R model for the Just a Line system. It will be useful to recapitulate the requirements for the new system, as these will to a large extent dictate what data we need to store (although not how we organize it).

- The system must produce a personalized combined price list and order form, to be sent out to existing customers. A non-personalized version of the same form will be distributed for publicity purposes.
- The system must check whether a customer order is valid, by checking that the card design ordered is one that is stocked. Checking that there is enough of that card design in stock to meet the order will remain a manual process. The computerized system must keep a record of outstanding customer orders.
- When the orders are made up, the system must produce invoices to go with the goods.
- The process of preparing a supplier order will remain manual, but the system must be able to list products and quantities on the current batch of outstanding customer orders, to facilitate supplier ordering.

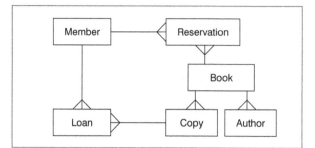

Figure 6.14 E-R model for the library system

E-R model for Just a Line

We will create this using the stages identified above.

Step 1. Identify entities

In the Just a Line system it seems a reasonable guess that we will want to store data about customers and their orders, and about the products the system deals with. Our initial list of candidate entities therefore is:

`Customer, Order, Product`

The system will certainly need to store details about these three entities. At this stage the selection of entities seems to be done by plucking them out of the air rather than anything more technical. In fact, all we are trying to do is achieve a starting position to which we can apply more technical methods.

Step 2. Attributes of Just a Line entities

Decisions about the attributes for these entities need some thought. At this stage it may be worth examining the data flow diagram of the current system in Figure 6.15. There are two data stores, Orders and ProductInformation. An examination of the contents of these data stores will be useful to us in the search for attributes for our entities. A simplified list of data store attributes is reproduced in Figure 6.16 in data modelling notation. We can use this as a starting point for the attributes required by the entities Customer and Product. Note that the inner parentheses around productDescription, packetSize, costPerPacket and costOfLine, mean that these are repeated attributes – each customer's order may be for many products.

Customer At the moment, the system stores no data about customers, except when an order is going through the system. However, we know that the new system must send mail shots to customers. A useful set of attributes, therefore, will be:

```
Customer
customerName
customerAddress
customerPhone
customerEmail
```

Product The current system has product information only on a scribbled price list, supplemented by the supplier's design list. The new system will need information about products for printing the price list + order form, for checking customer orders for validity, for pricing orders, for producing customer invoices and for producing the list of items on outstanding customer orders that is required to help with supplier ordering. The attributes we need to store about products, therefore, might be:

```
Product
productDescription
packetSize
costPerPacket
supplierName
supplierAddress
```

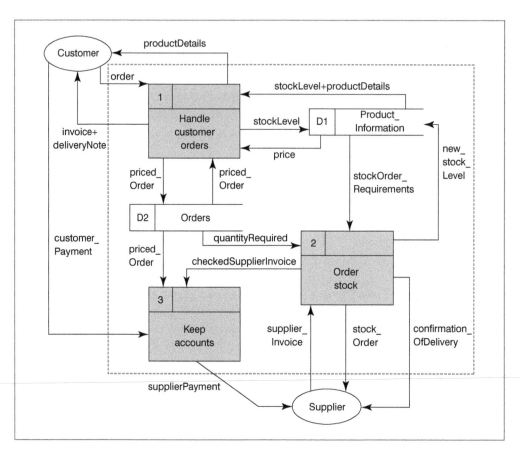

Figure 6.15 Just a Line: level 1 data flow diagram of the current system

Orders (customerName, customerAddress, customerPhone, customerEmail, dateRequired,
  (productDescription, packetSize, costPerPacket, costOfLine))

ProductInformation (productDescription, packetSize, costPerPacket, supplierName, supplierAddress)

Figure 6.16 Entity descriptions derived from the data stores ProductInformation and Orders

Order If we examine the current order form (Figure 6.17) we can see that orders in the current system include the following attributes (the list is simplified to make the example clearer).

Order
customerName
customerPhone
customerAddress
dateRequired

Just a Line

ORDER FORM

Mr/Mrs/Miss/Ms ______________________

Address ______________________

Delivery Address ______________________

Date Required ____________ Telephone ____________

Description	Packet Size	No. Packets	Price		£	p
			Total Cost			
			Delivery Charge			
			Amount Owing			

Figure 6.17 Just a Line order form

```
(productDescription²
packetSize
numberOfPackets
costPerPacket
costOfLine)
```

This list of attributes will serve as our starting point. Now we need to see what can be done to improve this initial list.

Steps 3, 4 and 5. Improving the E-R model

We now need to check that the data is in logical groups, identify primary keys and eliminate redundant data. We will do this for each of the three entities: Customer, Product and Order.

[2] productDescription, packetSize, numberOfPackets, CostPerPacket and costOfLine are repeated attributes.

Customer It is unlikely that customerName will always be sufficient to uniquely identify a customer. It will be safer to introduce a customer number (customer#). This gives us:

```
Customer
customer#
customerName
customerAddress
customerPhone
```

Product Cards are sold in packets of different sizes, packets of 5, 10, 25, etc. It will simplify identification of products if we introduce an internal productCode so that we do not need to rely on productDescription alone to identify individual products. If the same card is sold in two or more packetSizes then there will be a different productCode for each, so that the productCode identifies the card design and the number in the packet. This gives us:

```
Product
productCode
productDescription
packetSize
costPerPacket
supplierName
supplierAddress
```

It is unnecessary to store the supplier's name and address with every product – especially as Sue uses only one supplier at the moment. However, as she plans to use more than one supplier in the future, it would be useful to build in a reference to the supplier of each product. If we introduce a new entity Supplier and a unique number for each of Sue's suppliers, we can simply keep a reference to the appropriate supplier with the product details and put the supplier details in the new Supplier entity.

```
Product                 Supplier
productCode             supplier#
productDescription      supplierName
packetSize              supplierAddress
costPerPacket           supplierPhone
supplier#
```

Order At the moment our tentative list of attributes (see Figure 6.16) for Order is:

```
Order
customerName
customerPhone
customerAddress
dateRequired
```

```
(productDescription
packetSize
numberOfPackets
costPerPacket
costOfLine)
```

The first three items are already stored in Customer, and as we want to avoid storing redundant data they should not be repeated within Order. However, we do need to be able to link this order with the customer who placed it. Therefore, we can add customer# and omit the attributes customerName, customerPhone and customerAddress.

We need to be able to identify which products the customer has ordered. The safest way to do this is to include, for each item ordered, productCode, which uniquely identifies a product and the packet size. The attributes productDescription, packetSize and costPerPacket are stored in Product and can be accessed by using productCode; they are therefore redundant here; costOfLine can be calculated by multiplying numberOfPackets by costPerPacket – it can be derived from data we are storing anyway and can therefore be omitted. This gives us:

```
Customer
customer#
customerName
customerAddress
customerPhone
Product
productCode
productDescription
packetSize
costPerPacket
supplier#
Order
customer#
dateRequired
(productCode
numberOfPackets)
Supplier
supplier#
supplierName
supplierAddress
supplierPhone
```

The corresponding E-R diagram is shown in Figure 6.18.

Step 6. Investigating and recording relationships

In Figure 6.18 the relationship between Product and Order is modelled as many to many; one order may be for many products and one product may be on several orders. E-R modelling

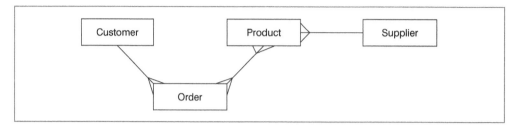

Figure 6.18 Just a Line – first cut E-R model

does not encourage many-to-many relationships. One reason for this is that there is often information associated with the relationship that cannot sensibly be attached to either of the participating entities.

To clarify this, it is worth digressing to consider a different example. An organization might want to record information about employees and ongoing projects. Employees can work on many projects and projects can be worked on by many employees. So we have a many-to-many relationship, as in Figure 6.19.

If the organization also wants to record the number of hours each employee works on each project, it is not obvious whether to keep this information in the Employee entity or the Project entity. If we record it in Employee, we will introduce redundancy into the model. What we will be trying to model can be recorded on paper as shown in Figure 6.20. However, E-R modelling presupposes that one record will be represented in one line of a table, which would mean that, unless we reorganize the data, this information would have to be recorded as in Figure 6.21, with an employee's number, name and address being repeated once for each project he or she works on.

The same repetition would happen if the information about hours worked were added to the Project entity. The information is associated with the relationship 'WorksOn'. The situation is resolved if we introduce an extra *intersection entity* to represent the information associated with the relationship, thereby splitting the many-to-many relationship into two, one-to-many, relationships (Figure 6.22). The data will now be recorded as in Figure 6.23.

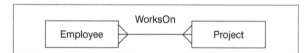

Figure 6.19 Many employees work on many projects

EmployeeNo	Name	Address	ProjectNo	HoursWorked
123	Smith J.	Hedgerley Close	4	23
			7	16
			3	3
456	Bloggs F.	West Road	6	1
			9	7

Figure 6.20 Hours worked per project, as represented on paper

EmployeeNo	Name	Address	ProjectNo	HoursWorked
123	Smith J.	Hedgerley Close	4	23
123	Smith J.	Hedgerley Close	7	16
123	Smith J.	Hedgerley Close	3	3

Figure 6.21 Records of hours worked in table form

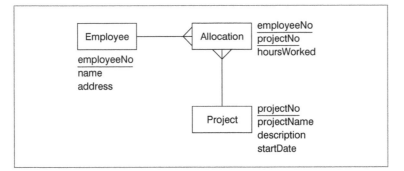

Figure 6.22 Intersection entity – Allocation

Employee	Project	HoursWorked
123	4	23
123	7	16
123	3	3
456	6	1
456	9	7

Figure 6.23 Data recorded in intersection entity Allocation

Many-to-many relationships can cause problems, even if there is no information associated with the relationship. The organization in our example will probably want to be able to link employees to projects as illustrated in the occurrence diagram in Figure 6.24 so that, given an employee number, they can list the projects he or she is working on and, given a project number, they can list the employees working on it.

We have the same problem – of deciding whether to keep the relevant employee numbers with the Project data or the project numbers with the Employee data. Either way, we will be

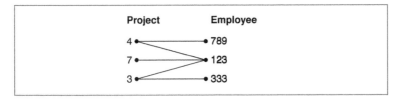

Figure 6.24 Occurrence diagram – project/employee

introducing redundancy of the sort illustrated in Figure 6.21. The solution again is to introduce an intersection entity to record the allocation of employees to projects (Figure 6.25). Another reason for not allowing many-to-many relationships is that most databases cannot implement them.

Returning to the Just a Line example, we can simplify the many-to-many relationship between Product and Order (Figure 6.18) into two, one-to-many, relationships, by introducing an intersection entity OrderLine. This gives us the diagram in Figure 6.26, which can be read as follows.

- Any one occurrence of the Order entity may have many OrderLines.
- Any one occurrence of the entity OrderLine is on only one Order.
- Any one occurrence of the entity Product may appear on many OrderLines.
- Any one occurrence of the entity OrderLine is for only one Product.

The attributes of Order and OrderLine are:

```
Order (customer#, dateRequired)
OrderLine (customer#, dateRequired, productCode, numberOfPackets)
```

Revisit step 4. Identify primary keys for Just a Line

Now that we have split the Order entity into two entities, we must define primary keys for the new entity OrderLine and the new version of the entity Order. So far, we have decided that the primary key for Customer will be customer# and the primary key for Product will be

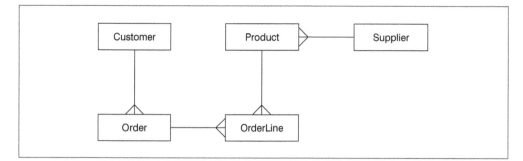

Employee	Project
123	4
123	7
123	3
456	6
456	9

Figure 6.25 Intersection entity, records allocation of employees to projects

Figure 6.26 Simplified E-R model

productCode. For each order to be uniquely identifiable we could either introduce order numbers (as we do in our Access implementation of Just a Line in Chapter 8) or make a decision that the customer# and dateRequired, taken together, will uniquely identify any order (as we do here). If we assume one product per OrderLine, then each OrderLine can be identified by a combination of the attributes customer#, dateRequired and productCode.

Summary of attributes
(Key attributes are underlined.)

```
Customer (customer#, customerName, customerAddress, customerPhone,
customerEmail)
Product (productCode, productDescription, packetSize, costPerPacket,
supplier#)
Order (customer#, dateRequired)
OrderLine (customer#, dateRequired, productCode, numberOfPackets)
Supplier (supplier#, supplierName, supplierAddress, supplierPhone)
```

What have we achieved by building the E-R model?
The E-R model should have captured all the data required by the system (and no data in excess of these requirements) and the relationships between groups of data items. The process of creating this model and documenting it forces us to study the system data in detail. The model provides a basis for useful discussions with the users. However, E-R modelling has a rather informal approach to the identification and structuring of the data. The technique is intuitive and subjective: we identify the entities as the things about which we think the system needs to collect data. It is easy to miss entities. We allocate attributes to the entities where they seem (to us) to belong, and build in relationships we perceive to be useful. Normalization, discussed below, provides a more formal and rigorous approach to the identification of entities, their attributes and the relationships between them. It also provides a useful check on the E-R model.

6.3 Normalization
Introduction
To reiterate, the aims of data modelling are as follows.

- The identification of the data objects or entities in the system, their structure and the relationships between entities.
- The construction of a model of the stored data requirements of the system that is independent of specific processing requirements.
- The construction of a robust data model, i.e. of a minimal model of the data required to be stored by the system.
- The construction of a logical model of the data, i.e. a model that is not concerned with how the data storage will be, or is currently, physically implemented.

E-R modelling sets out to achieve these objectives using a top-down approach and the informal application of a set of guidelines. Normalization provides an algorithm for reducing complex data structures to irreducible simple structures. It has been observed over the years that some data groupings behave better in an insert/update/delete situation. This observation has been formalized in a set of rules known as Codd's laws. During normalization we rigorously apply these rules to the mass of data accumulated during the fact-finding stage of the analysis of the system. Once we have produced normalized entities, the process of normalization provides automatic mechanisms for mapping the relationships between entities.

Although normalization gives us a set of rules for determining whether we have grouped our data items correctly, it cannot, in itself, ensure that we have captured all the data the system may require, or that we have fully understood it. If we do not have a sound understanding of the data and relationships between data items in a system, normalization will not magically produce a good data model for us. This is one reason why it is useful to use a two-pronged approach: E-R modelling and normalization.

Normal form

The term **normal form**, when applied to data, simply means a convenient form or structure into which the data can be organized – in this case the forms dictated by Codd. We shall concern ourselves with the first, second and third normal forms. In fact, several higher forms have been defined, but these are beyond the scope of this book. A data model normalized to the third normal form is useful, however we decide to implement, in that it will have removed data redundancy and grouped the data logically.

Determinancy and dependency

The concepts of dependency and determinancy are both important in the context of normalization. The terms express complementary concepts – if A determines B, then B is dependent on A. A determines B if each value of A is always associated with only one value of B. For example, in the dentists' appointment system shown in Figure 6.27, if every time we come across the patient name John Knowles we find that the associated dentist is Ian Paterson, and every time we see the patient name Jane Hamilton, the dentist is Chris Dean, this shows us that, in this data, the patient name determines the dentist and, conversely, the dentist is dependent on the patient name. In other words, if we know the patient name, we will be able to find the associated dentist. By contrast, a dentist's name can be associated with more than one patient: Chris Dean is associated with both Jane Hamilton and Jennifer Seath; Ian Paterson has John Knowles, Jill Crerar and Sam Ollis as his patients. The dentist's name therefore, does not determine the patient.

An example of normalization

To illustrate the process of ensuring that data is in the first, then second, then third normal form, a simple example will be used. The situation we shall consider is the data recorded by a firm of civil engineers on its staff development course form, illustrated in Figure 6.28. The entity StaffDevelopmentCourse has the following attributes:

```
StaffDevelopmentCourse
courseCode
courseDescription
(employee#
name
block
room#
```

Patient No.	Patient Name	Appointment Date	Appointment time	Dentist
2555	Jane Hamilton	12/3/05	11.30	Chris Dean
2534	Jennifer Seath	12/3/05	11.30	Chris Dean
3216	Jill Crerar	12/3/05	11.45	Ian Paterson
5632	John Knowles	13/3/05	15.00	Ian Paterson
2555	Jane Hamilton	13/3/05	15.00	Chris Dean
5632	John Knowles	15/3/05	10.20	Ian Paterson
2555	Jane Hamilton	15/3/05	10.30	Chris Dean
2397	Sam Ollis	26/4/05	09.40	Ian Paterson

Figure 6.27 **Extract from a dental practice's appointment system**

STAFF DEVELOPMENT COURSE					
COURSE CODE: *FRLANGS*					
COURSE DESCRIPTION: *Conversation*					
Employee Number	Name	Block	Room	Joined Course	Allocated Time (hours)
213	Jones H.N.	J	124	5/4/05	24
164	Smith J.E.	H	603	1/5/05	6
465	Baker K.P.	G	21	12/7/05	12
324	Paterson W.I.	H	603	9/7/05	10

Figure 6.28 **Staff development course form**

```
dateJoinedCourse
allocatedHours)
```

In line with recent company policy all staff are released for two hours a week to improve their foreign language skills by following tapes in the laboratory. This form records data about foreign language courses, a particular course being identifiable by its courseCode (which has therefore been underlined as a primary key). Several employees can be working on the same course (they work at their own pace using the language laboratory) and an employee may be working on more than one course. A total of six attributes is recorded about each employee, including their normal office location (block and room number), the date they joined this course and the number of hours they have been allocated to the course. Employees may share the same office.

We can immediately see that this does not form a satisfactory entity. For example, much of the employee information in StaffDevelopmentCourse may be duplicated in employee records elsewhere; if it is not, all record of an employee vanishes for any period when that person is not working on a course.

First normal form (1NF)

Definition An entity is in 1NF if, and only if, it has an identifying key and there are no repeating attributes or groups of attributes.

Therefore, to get into 1NF we must remove the repeating group:

■ the attributes that do not repeat are left as one entity, now a 1NF entity
■ the repeating attributes are removed into a separate 1NF entity, which includes the key (courseCode) of the original entity to provide the link between the two tables, and a new key (employee#) for the repeating group of attributes.

Applying this rule gives the following two 1NF entities:

Course	EmpOnCourse
courseCode	courseCode
courseDescription	employee#
	name
	block
	room#
	dateJoinedCourse
	allocatedHours

The entity EmpOnCourse has a combined attribute primary key: courseCode, employee#. The problem of vanishing employees has not yet been solved; the second normal form will help us with this.

Second normal form (2NF)

Definition An entity is in 2NF if, and only if, it is in 1NF and has no attributes that require only part of the key to identify them uniquely. Therefore, to get into 2NF we remove part-key dependencies:

■ where a key has more than one attribute, check that each non-key attribute depends upon the whole key to determine it, and not just part of the key
■ for each sub-set of a key that identifies an attribute or set of attributes, create a new separate entity.

This sounds daunting, but all that is really happening is that the data is being split into logical groups of data items that go together. Notice that we need only apply the 2NF rule to entities with multiple-attribute keys; it does not apply to entities with single-attribute keys.

In our example, Course, having a single-attribute key, is already in the second normal form. However, EmpOnCourse is not, and requires some consideration. As noted above, courseCode combined with employee# forms the primary key for this group of attributes. The following list shows which attributes depend on just part of this key, and which depend on the whole of this key.

Attribute	Depends on
name	employee#
block	employee#
room#	employee#

but

dateJoinedCourse	courseCode + employee#
allocatedHours	courseCode + employee#

If you are in any doubt about this ask yourself: are employees' names affected when they join a course? The answer is obviously not. Therefore, name is not dependent on courseCode, only on employee#. We could imagine a situation where an employee's room number and block number change when he or she starts a course: this would be the case if we were recording employees' current locations. However, in our example, the room and block numbers refer to employees' permanent office numbers and are not affected when they start a course. The date in question is the date that one particular employee joined one particular course and therefore is dependent on both the employee number and the course code. The same is true of the number of hours an employee has been allocated to a particular course.

For the attributes that depend on only part of the key we must create a separate entity keyed by the attribute(s) they depend on. So name, block and room# must be moved into a new entity, Employee, keyed by employee#; dateJoinedCouse and allocatedHours remain in the original entity keyed by the joint-attribute key <u>courseCode,employee#</u>.

This gives us three 2NF entities:

Course	EmpOnCourse	Employee
<u>courseCode</u>	<u>courseCode</u>	<u>employee#</u>
courseDescription	<u>employee#</u>	name
	dateJoinedCourse	block
	allocatedHours	room#

This is better. We no longer lose employees who leave courses as we have a separate entity for employee data that will be in place whether or not an employee is registered on a course. But there is still a problem. Presumably, block and room# are related in some way; if one is updated, the other may be affected. This again is a question about the meaning of the data. In this case the company is located on a site with several separate buildings or blocks. Each block contains several rooms, as illustrated in Figure 6.29, which shows a few examples of how room numbers relate to blocks. Although a room# does uniquely identify a room, the company has not assigned room numbers to its offices in a systematic way.

There is a problem with storing both block and room# in Employee with the rest of the data about the employee. Rooms may be shared by employees – for example, 50 employees may share the same large open-plan office, room# 18. The fact that this room is in block G will be

block	room#	block	room#	block	room#	block	room#
G	15	H	19	I	1	J	7
G	16	H	20	I	2	J	9
G	17	H	21	I	3	J	11
G	18	...		...		...	
...		...		...		...	

Figure 6.29 Block and room numbers

repeated 50 times in the Employee entity. This wastes space and allows errors to creep in; we might find that a room number has erroneously been recorded as being in two different blocks.

Third normal form (3NF)

Definition An entity is in 3NF if, and only if, it is in 2NF and no non-key attribute depends on any other non-key attribute.

Therefore, to get into 3NF we must remove attributes that depend on other non-key attributes:

- decide on the direction of the dependency between the attributes
- if A depends on B, create a new entity, keyed by B, with A as an attribute (A may be a set of attributes, B may be a compound key)
- leave B in the original entity and mark it as a foreign key, but remove A from the original entity.

To work out whether A is dependent on B we must discover whether, given a value for B, there is only one possible value for A. If so, A is dependent on B. Relating this to our example, we can work out from the situation outlined in Figure 6.29 that, given any value of block, there is more than one corresponding value for room#, since each block contains many rooms. Therefore, room# is not dependent on block. However, for any given room#, there is only ever one possible corresponding value for block. Therefore, block is dependent on room#. We therefore create a new entity keyed by room# with block as an attribute, and remove block from the original Employee entity. Employee (2NF) gives two 3NF entities:

```
Employee (2NF)        becomes        Employee (3NF)
employee#                            employee#
name                  name
block                 room#*
room#

                                     and

                                     Location
                                     room#
                                     block
```

A foreign key is marked by an asterisk; in this case room#*. Identifying an attribute as a foreign key means that an attribute in one entity is the primary key of another entity (see the Terminology and notation subsection in Section 6.1). When we look at the full set of entities below, we can see that courseCode* and employee#* (in EmpOnCourse) are also foreign keys and are marked as such.

The total list of 3NF entities is:

```
Course (courseCode, courseDescription)
EmpOnCourse (courseCode*, employee#*, dateJoinedCourse,
             allocatedHours)
Employee (employee#, name, room#*)
Location (room#, block)
```

The problems caused by storing room# and block in Employee have now been resolved. We only record once which block a room is in. There is no longer any risk that the same room number could be recorded as being in more than one block.

It is useful to represent these normalized entities and their relationships diagrammatically, using the same conventions as in E-R modelling. Relationships can be worked out automatically using the following rules.

1. Given two entities A and B, if the key of A is a sub-set of the key of B then the relationship of A to B is one to many (Figure 6.30a).
2. Given two entities P and Q, if P contains a foreign key F and F is the primary key of Q, then the relationship of P to Q is many to one (Figure 6.30b).

Figure 6.30a The key of A is a sub-set of the key of B

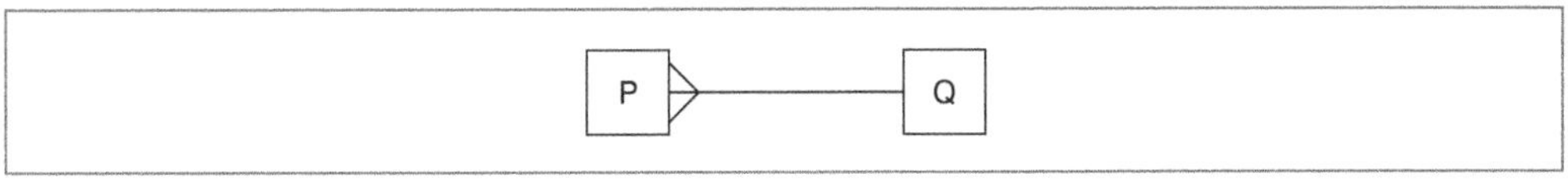

Figure 6.30b The key of Q is a foreign key of P

To apply these rules we need to be clear about the keys to the entities. In our example we have the situation illustrated in Figure 6.31.

■ The key of Course (<u>courseCode</u>) is part of the compound key of EmpOnCourse (<u>courseCode, employee#</u>). Rule 1 applies; therefore the relationship between Course and EmpOnCourse is one to many.

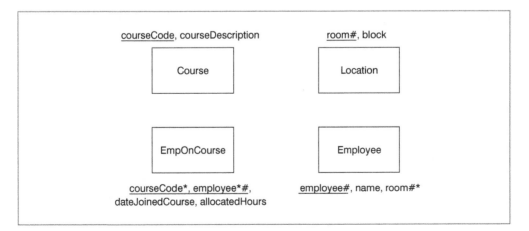

Figure 6.31 Staff Development Course entities and attributes

- Similarly, the key of Employee (employee#) is part of the compound key of EmpOnCourse (courseCode, employee#). Rule 1 applies; therefore the relationship between Employee and EmpOnCourse is one to many.
- The entity Employee has a foreign key (room#), which is the single key of the entity Location. Rule 2 applies; therefore the relationship between Employee and Location is many to one.

The completed diagram in Figure 6.32 shows the relationships between the entities.

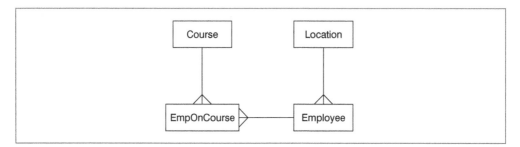

Figure 6.32 E-R diagram of Staff Development Course example

What we have achieved once normalization is complete

Normalization provides a useful and rigorous check on the less formal activity of E-R modelling. The recommended approach to data modelling is to adopt the intuitive E-R modelling approach first, then use normalization to check that the entities produced are logical data groups, with no redundant data. We suggest that you check that the Just a Line entities listed above are normalized.

When working on large systems where we are applying normalization techniques to the data from many documents and files, it is often the case that we end up with several sets of entities, from different sources. The sets of entities will need to be rationalized: duplicated entities can be deleted, entities that have the same key but different or overlapping sets of attributes can be combined. Sometimes we may decide to replace a multiple-attribute key produced during the normalization process with a single primary key that is easier to work with. When we implement the Just a Line system in Access (see Chapter 8), we replace the multiple-attribute key of Order (customer#, dateRequired) with the simpler key, order#.

Summary

Data modelling uses two techniques: a top-down approach, E-R modelling; and a bottom-up approach, normalization. E-R modelling starts by intuitively identifying objects (entities) about which the system is required to store data. Appropriate attributes are then allocated to each entity. Normalization starts with the smallest meaningful data items in the system and organizes them into well-formed entities. Normalization is a useful check that the entities identified during E-R modelling are logical data groups and that there is no redundant data.

Exercises and topics for discussion
What can you remember?

You will find all the answers in the chapter

a) What is the basic difference in approach between E-R modelling and normalization?

b) What do the following terms mean?
- Entity
- Occurrence
- Attribute
- Primary key
- Foreign key

c) What is the difference between an optional and a mandatory relationship?

d) Give an example of a many-to-many relationship; draw a diagram to show the relationship.

e) Why does E-R modelling not encourage many-to-many relationships?

f) What do we mean by redundant data?

g) What is the aim of normalization?

h) What are determinancy and dependency?

i) What is first normal form?

j) What is second normal form?

k) What is third normal form?

6.1 (a) From the information presented in Figure 6.2 in this chapter, list the candidate keys.

(b) State which of the candidate keys would make the best primary key. Give reasons for your choice of primary key and justify your rejection of the other candidate keys.

6.2 Draw E-R diagrams of the following situations.

(a) A tree has many leaves; a leaf is on only one tree.

(b) A manager manages one laundrette; a laundrette is managed by only one manager.

(c) A car has one car owner; a car owner may own several cars.

(d) A hockey team has many players; a player may only play for one team.

(e) An author writes many books; a book may be written by several authors.

(f) A father may have several children; a child has only one father.

6.3 Examine the E-R diagram in Figure 6.33 and state which of the following statements are true.

(a) A course can only be on one scheme.

(b) A scheme has only one course.

(c) A student does not have to be on any scheme.

(d) A student must be on one scheme only.

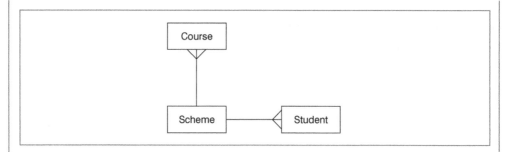

Figure 6.33 E-R diagram of student course example

6.4 In your group medical practice each doctor has many patients but each patient may be registered with only one doctor. Assume that a doctor can be identified by name and a patient by patientNo.

(a) Draw an E-R diagram to model this situation.

(b) How does the model change if patients are allowed to be registered with more than one GP?

6.5 Simplify the many-to-many relationship shown in Figure 6.34 by introducing an intersection entity. Find a name for the new entity.

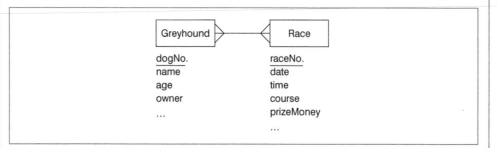

Figure 6.34 Many-to-many relationship

6.6 Fenland College of Knowledge wishes to record information about students and their assignments. A fragment of the current manual recording sheet is shown in Figure 6.35. Student numbers uniquely identify students, subjects are identified by subject code, and assignments are identified by subject code and assignment number.

Student No.	Name	Subject	Lecturer	Assignment	Description	Grade
32	J. Crerar	IS	C.L	1	E-R model	B
		SML	C.R.P	2	clock program	C
45	I. Seath	C++	S.B	3	lift program	C

Figure 6.35 Fragment of recording sheet concerning students and their assignments

(a) Write entity descriptions for the entities Student, Subject and Assignment, underlining keys.

(b) Draw an E-R diagram showing relationships between these entities.

Exercise 6.6 is a much more difficult exercise than the previous ones – do not be discouraged if you find it hard.

6.7 The data below shows exam marks and their equivalent grades. In this example does the grade depend on the mark, or the mark depend on the grade?

Mark	Grade
80–100	A
65–79	B
50–64	C
40–49	D
25–39	E

6.8 Each lecturer in a department is personal tutor to a group of first-year students. Examples of two groups are shown below. In this case does the lecturer depend on the student, or the student on the lecturer?

Lecturer	Student
Mark Jones	Naresh Patel
	Chris Smith
	Sally Holmes
	Jo Ann Lane
Nicky Hammond	Garry Lee
	Harry Bing
	Rita Fraser
	Susie Franks

6.9 A video shop keeps records of its loans to customers as shown below:

```
Customer (customer#, name, address, (videoRef.,
dateBorrowed, dateDueBack))
```

Put the data into first normal form.

6.10 A company wants to keep a record of the languages spoken by its employees, as shown below.

```
Employee (employee#, name, department, (language, level))
```

Put the data into first normal form.

6.11 A sports teacher in a school wants to record pupils' running times, as shown below.

```
RunningTime (pupilName, class, height, weight, (distance,
time))
```

Put the data into first normal form.

6.12 A university department records students' final-year options as shown below.

```
FinalYearOptions (student#, studentName, (optionCode,
optionName, lecturer, studentGrade))
```

Put the data into first and second normal forms.

6.13 The unnormalized data for a book loan is shown below. Put the data into first and second normal forms.

```
BookLoan (borrower#, borrowerName, borrowerAddress,
(book#, title, author, ISBN, dateBorrowed, dateDueBack))
```

6.14 The unnormalized data for hotel room bookings is shown below. Put the data into first and second normal forms.

```
RoomBookings (room#, roomType, roomPrice, (guestName,
guestAddress, dateOfArrival, dateOfDeparture))
```

6.15 A junior school teacher wants to keep track of how her pupils are progressing with the class readers, as shown below. Put the data into first and second normal forms.

```
PupilReaders (pupilName, dateOfBirth, (reader#,
readerTitle, vocabularyLevel))
```

6.16 A doctors' surgery wants to keep records of the referrals made by the various doctors in the practice, as shown below. Put the data into first, second and third normal forms. Identify any foreign keys.

```
Referrals (doctorCode, doctorName, (patientName,
patientAddress, patientPhone#, diagnosisName,
dateOfReferral, consultant, hospital))
```

6.17 A university department wants to record details of multiple-choice tests, as shown below.

```
MultipleChoiceTestbank (testID, topic, level,
estimatedTimeToComplete, authorName, department, email,
(question#, questionType))
```

Put the data into first, second and third normal forms. Identify any foreign keys.

6.18 A hospital consultant's list contains the data below. Assume that a consultant will be in only one unit on any one day. A consultant is uniquely identified by his or her consultant number, a patient by his or her patient number, and a GP by his or her GP number.

```
Consultant'sList (consultant#, consultantName,
consultantPhone#, (day, unit, (time, patient#,
patientName, patientAddr, GP#, GPName, GPAddr, GPPhone#)))
```

Put the data into first, second and third normal forms.

6.19 A company sells its kitchenware products by means of a team of salespersons who organize kitchenware parties in their customers' houses. Products are supplied to the customers from the nearest warehouse. A salesperson may have several warehouses in his or her sales area. The company keeps records of the total value of sales per customer for each salesperson, as shown in Figure 6.36.
Describe this data in unnormalized form, in first normal form, in second normal form and in third normal form.

Salesperson Number: 369				
Name: Webster				
Sales Area: South-east				
Customer#	Customer Name	Warehouse#	Warehouse Site	Total Sales (£)
46732	Bailey	34	Landbeach	453.56
26777	Sadler	23	Cottenham	205.03
23331	Bloxham	12	Waterbeach	34.99

Figure 6.36 Record of the total value of sales per customer for each salesperson

6.20 An animal shelter keeps a record of the treatment prescribed to each animal in its care. Details are recorded on a treatment card; a typical example is shown in Figure 6.37. Animals are uniquely identified by their number. The shelter is divided into blocks of animal cages: one block for dogs, one for cats and one for animals with infectious diseases. Each block has a unique number. Describe the data on the card in unnormalized form, in first normal form, in second normal form and in third normal form.

NAME: Fred BLOCK NO: 6		BREED: Collie cross TYPE: IsolationBlock		NUMBER: 87 BLOCK NAME: Tara	
Start date	**Condition**	**Drug code**	**Drug name**	**Dosage**	**Days**
31/8/05	bite on right hind leg	123	Trib	1 × 80 mg twice a day	10
31/8/05	infection	234	Metron	3 × 200 mg once a day	8
6/9/05	fleas	567	Nuvan Top	once every 10 days	

Figure 6.37 An example of a treatment card

6.21 The People's Prose Book Club operates a mail-order service. Members select books from titles on offer for any particular month. The club keeps track of member details, and the books they have ordered, on a membership record card. An example of a membership record card is shown in Figure 6.38. Membership numbers, title numbers and publisher codes are allocated by the book club. Describe this data in unnormalized, first, second and third normal forms.

Member Number	Name		Address	
452	Ms E. Gorse		8 The Beeches, Waterbeach	
Date Ordered	**Title Number**	**Title**	**Publisher Code**	**Publisher Name**
2/4/05	234	Plato and Pleasure	86	Finn
2/4/05	145	Loaves and Fishes	23	Browne
14/5/05	136	Picnics for Fun	23	Browne

Figure 6.38 Example of a membership record card

Further reading

Date, C.J. (2003) *An Introduction to Database Systems*, 8th edn, Addison-Wesley, Reading, MA.

Goodland, M. and Slater, C. (1995) *SSADM Version 4: A Practical Approach*, McGraw-Hill, London.

Hawryszkiewycz, I. (2001) *Introduction to Systems Analysis and Design*, 5th edn, Prentice-Hall, a division of Pearson Education, Australia.

Hoffer, J.A., George, J.F. and Valacich, J.S. (2004) *Modern Systems Analysis and Design*, 4th edn, Prentice-Hall, a division of Pearson Education, NJ.

Howe, D.R. (2001) *Data Analysis for Database Design*, 3rd edn, Butterworth-Heinemann, Oxford.

Kendall, E.K. and Kendall, J.E. (2005) *Systems Analysis and Design*, 6th edn, Prentice-Hall, a division of Pearson.

CHAPTER 7
EVENT
MODELLING

In this chapter we look at:

- the views that different modelling techniques give of the system

- the role of entity life histories and state diagrams

- how to draw an entity life history

- how to draw a state diagram

After studying this chapter and working through the exercises you will be able to:

- compare the views that different modelling techniques give of the system

- explain the role of entity life histories and state diagrams

- draw a simple entity life history

- draw a simple state diagram

Introduction

Each of the techniques that we have met in Chapters 4, 5 and 6 provides a different view of the problem area and the evolving system. Data flow diagrams, for example, give us a picture of the movement of data through the system at various levels of detail. It is tempting to read a data flow diagram from left to right and from top to bottom, and to assume that the process numbers indicate the order in which things happen in the system, but this would not be accurate. Data flow diagrams do not include considerations about the sequencing, iteration or timing of events in the system. It is not part of a data flow diagram to describe the order in which processes operate on data, nor how many times a particular process may need to be carried out.[1]

To illustrate this point, we can look at Figure 7.1, which shows part of the current level 1 data flow diagram for the Just a Line system.

The diagram does not tell us whether the customer places an order before receiving details of the cards, or whether Harry and Sue send product details to potential customers in the hope that they will then place an order. The data flow diagram is equally vague about when a payment is made. We have no way of knowing whether all orders must be paid for in advance, whether Harry and Sue operate a cash on delivery system or whether credit is available.

The important point is that this lack of detail about the order of events in the system is not a fault in the data flow diagram. As we said at the beginning of Chapter 4, all models represent

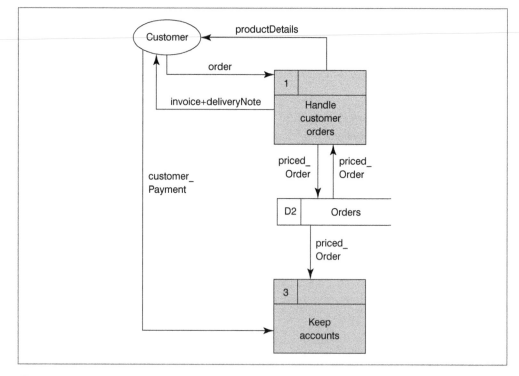

Figure 7.1 Part of the level 1 data flow diagram for the Just a Line system

---

[1] More advanced data flow diagrams, such as those used in the development of real-time systems, do model sequence and flow of control, but these are beyond the scope of this book.

an abstraction, or view of the system. It is not part of the role of a data flow diagram to model sequence, timing or iteration in the system. Nor is this the role of the data model, which simply provides a view of the relationships between data objects. The data model ignores the system processes, and so considerations of sequencing and iteration are not included.

It is possible to introduce the concepts of sequence and iteration into process definitions by means of *if ... then* statements and *while ... do* loops, but these definitions are used only to describe the lowest-level processes on a data flow diagram. They do not give a picture of the system as a whole.

To obtain a view of the ordering of events in a system, and the way in which entities respond to external events, we are going to look at two techniques: **entity life histories** and **state diagrams**. Entity life histories have been around for a long time, and are less frequently used than they were; however, they are still included in some mainstream methodologies, such as SSADM (Goodland and Slater, 1995), and that is why we discuss them here. State diagrams have been used for many years in **real-time system** development, and are increasingly widely used to model the ways that any system can respond to external **events**. There are a number of variations on the technique and a number of different names – for example, you will see these diagrams referred to as state charts and state transition diagrams, but the fundamentals explained in this chapter will enable you to cope with the alternatives.

Whether entity life histories or state diagrams are used, the models are generally produced after data and process modelling. This is because these techniques use the entities and processes identified in the E-R and data flow diagrams; they combine the static view of the system provided by the data model with dynamic view of the system provided by the data flow diagrams.

7.1 Entity life histories

Entity life histories (ELHs) are a diagrammatic technique that provides a picture of all possible biographies for any occurrence of a particular entity in the system. ELHs model all the events that can affect the stored data in the system and show what effect they can have. For example, in the current (manual) Just a Line system, the events in the life of a customer entity are as follows.

- A customer places an order – this will cause a customer entity to be created.
- The customer's details (such as the address) change during the life of the order – this will cause the entity to be amended.
- The order is delivered or cancelled – this will cause the entity to be deleted.

The entity life history for a customer in the current Just a Line system can be seen in Figure 7.2.

The notation for the diagram is explained in detail later in the chapter. The diagram gives us a picture of the events listed above. It is interesting in that it shows that Harry and Sue have no way of keeping a record of customers, except for those who have an order currently being handled by the system. Because they give a different perspective on the system, that of ordering and timing of events, entity life histories often highlight interesting details that may be missed in other structured models.

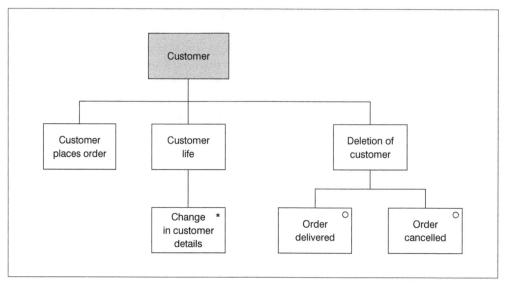

Figure 7.2 Entity life history of a customer in the current Just a Line system

Some definitions relating to entity life histories

■ An entity is a data object of interest that has been identified in the entity-relationship model. In the Just a Line system, customers, orders and cards are all objects of interest, giving us the entities Customer, Order and Product.

■ An event is something that happens in the real world, which causes an entity (or more than one entity) to be updated. For example, a customer may move house or the price of a product may change. These are both events that will respectively affect the entities Customer and Product.

Notation for entity life histories

Entity life history (ELH) notation provides constructs to express sequence, selection and iteration of events in the system. Figure 7.3 illustrates sequencing. The diagram is read from left to right, so that event 1 in the diagram must take place before event 2, and event 2 before event 3.

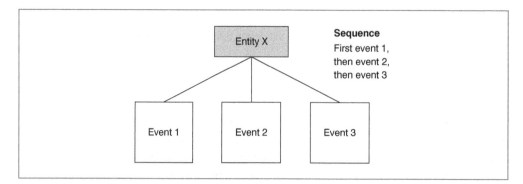

Figure 7.3 Representing sequence in entity life histories

Figure 7.4 shows how iteration (or repetition) is represented in entity life histories. The asterisk, *, in the event 2 box shows that, for a specific instance of the entity, this event may occur any number of times, or it may not occur at all. The blank box in this diagram is part of the entity life history structure. Such boxes are usually given an appropriate label (see the two example ELHs in Figures 7.8 and 7.9).

Alternative events are shown in Figure 7.5. In this diagram, the 'o' symbol shows that either event 2 or event 3 will take place, but not both. Figure 7.6 is very similar, but in this diagram event 2 may occur or it may not. There is no alternative event.

Although entity life histories are read from left to right, there are a few cases where this imposes an unnecessary constraint on the ordering of events in a system. An example of this

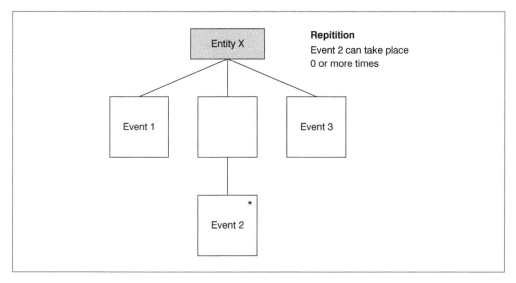

Figure 7.4 Representing iteration (repetition) in entity life histories

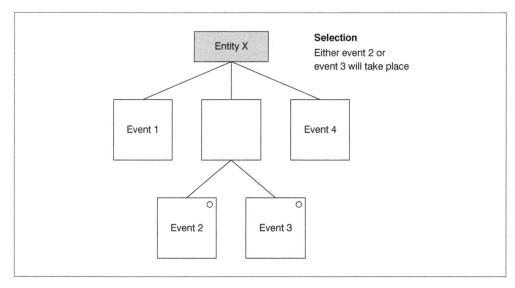

Figure 7.5 Representing selection in entity life histories

127

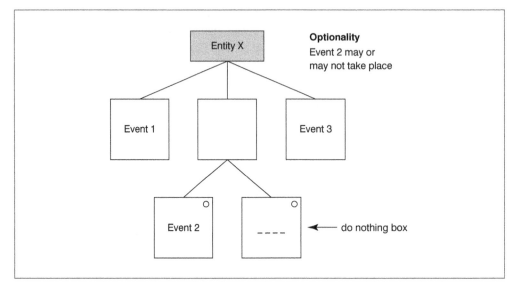

Figure 7.6 Representing optionality in entity life histories

from Just a Line would occur if it were possible for customers to pay for their cards either when they order (before they receive the invoice) or when the goods are delivered with the invoice. In this case we want to show that different orderings of the 'Send invoice' and 'Customer payment' events are possible. Figure 7.7 illustrates the double-line notation, which indicates that certain events can take place in any order.

It is important to remember that different types of symbol (*, o, etc.) cannot occur at the same level from the same box. To preserve the structure of the entity life history we use blank boxes (known as *structure boxes*), as in Figure 7.4.

How to create an entity life history

The first step is to identify all the events that affect the system data, and note whether their effect on each entity is to action one of the following:

■ create a new occurrence (e.g. when a customer places an order this will create an occurrence of the entity Order)

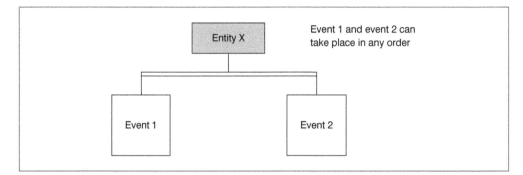

Figure 7.7 The double line indicates that there is no restriction on the ordering of events 1 and 2

- update an occurrence (e.g. if details of the order are altered)
- delete an occurrence (e.g. when the details of an order are removed from the company's records).

It is important not to confuse events with the processing that they initiate. It may take more than one event to trigger a process; for example, a particular product line may be sold out and it is decided not to stock the line any more. These two events together trigger the process of deleting details of this particular product. On the other hand, the company may decide to discontinue the product and simply write off the unsold items. In this case the decision to discontinue the product is the only event needed to trigger the deletion process.

In entity life histories we do not normally record events that have no effect on the **state** of an entity. For example, if a customer wants to know which cards he or she has ordered, this enquiry will not affect the Order or the Customer entities.

Since each occurrence of an entity must be created, and eventually deleted, we must check that events exist to trigger creation and deletion. If no deletion-triggering event exists, we must create a system housekeeping event to prevent the file growing in an uncontrolled manner.

Example ELH from Just a Line – the Product entity life history

We will assume a data dictionary definition of the entity Product as:

```
Product = productCode + productDescription + packetSize +
          costPerPacket
```

We now need to identify all the events that may affect Product and decide whether each event creates, amends or deletes an occurrence of the entity. This can be done by means of an entity-event matrix:

Event	Effect on Product
new product line stocked	create
price increase on product	amend
different number of cards per packet	amend
product becomes best-selling line	no effect
decision to discontinue product line	delete

Figure 7.8 shows the life history of the Product entity. An occurrence of the entity is created when a new line is stocked. This is followed by the Product life, which may consist of any number of amendments, including zero. Each amendment may be either a change in price or a different number of cards in a packet. Finally, the occurrence of the Product entity is deleted when the line is discontinued.

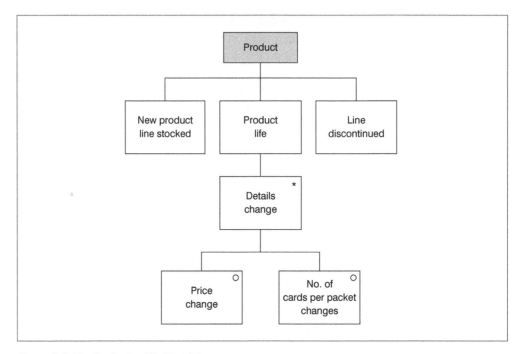

Figure 7.8 The Product entity life history

A second example ELH from Just a Line – the Order entity life history

We will assume a simplified data dictionary definition of the Order entity:

```
Order = customerNo + dateRequired + {orderLine}
```

Event	Effect on Order
customer places an order	create
order is validated	amend
order details are changed	amend
order is cancelled	amend
order is delivered	amend
one month after completion of order (housekeeping)	delete

It may seem strange that the events of cancelling and delivering an order amend the entity instead of deleting it. This is because Harry and Sue wish to keep records of past orders for one month. Notice that although the data model views Order and OrderLine as two entities, for the purpose of the entity life history and state diagram, they are regarded as a single entity unless there is a good reason to model them separately – if, for example, they are affected by a different set of events. You can see the Order entity life history in Figure 7.9.

Status indicators

Numbers can be added to the event boxes on an ELH to show which state the entity is in at any particular time, and place restrictions on the sequence in which events can happen. These numbers are known as status indicators. Figure 7.10 shows a more detailed version of the

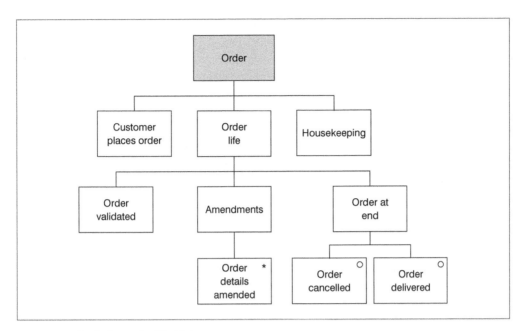

Figure 7.9 The Order entity life history

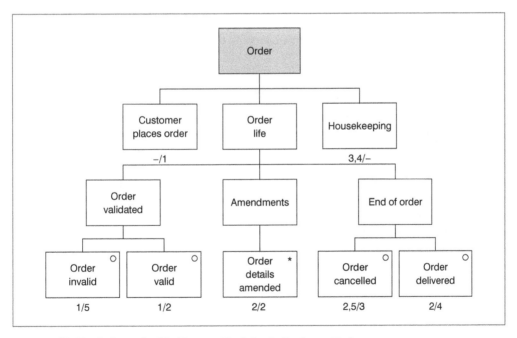

Figure 7.10 The Order entity life history with status indicators added

Order entity life history with status indicators added. Exercise 7.1 at the end of the chapter tests your understanding of the diagram.

The status indicators give us the following extra information about the Order entity.

■ When a customer places an order the occurrence of the entity Order moves into state 1 (it has been created, but nothing more).

■ When validation of the order takes place one of two things may happen. Either the occurrence of the Order entity is invalid, in which case it goes into state 5, or the order is valid and so moves into state 2. Validation can occur only when the order has been created and is in state 1. This is expressed by the status indicators 1/5 (which appears under the Order invalid event box) and 1/2, meaning a transition from state 1 to state 2 (which appears under the Order valid event box).

■ The status indicator 2/2, which appears under the Order details amended event box, means that an amendment to an order is not an event that changes the state of the entity. It also tells us that order details can be amended only once the order has been validated (i.e. it is in state 2).

■ The status indicator 2,5/3, defines the conditions under which an order may be cancelled. The order may have been placed but is invalid (state 5), it may have been placed and validated (state 2), or it may have been placed, validated and subsequently amended (state 2); in each of these cases it can be cancelled and so moves into state 3.

■ An order may be delivered when it has been validated or when it has been validated and amended. It may not be delivered when it has simply been placed or if it is found to be invalid. This is shown by the status indicator 2/4 under the Order delivered box, which moves the order into state 4.

■ Finally, the status indicator under the housekeeping box 3,4/– shows that an occurrence of the Order entity can be deleted only after it has been cancelled, or when it has been delivered. Once an order has been placed, it must go through the validation process and cannot be deleted until it has been either cancelled or delivered.

Status indicators are also useful as reference points for other modelling techniques. Process definitions often refer to the state of an entity – for example ,'If Customer is in state 2 ...'.

7.2 State diagrams

State diagrams are an increasingly popular technique for modelling the ways in which the entities in a system behave during their lifetimes and how they respond to external events. A state diagram describes how an entity – for example, Customer or Product in the Just a Line system – behaves in the system; in other words, how it reacts to all the external events that affect it. All the instances of the entity (the individual customers or products) have the same range of ways in which they can behave, but the actual way an individual instance does behave during the running of the system depends on the sequence of events that it experiences.

Some definitions relating to state diagrams

You will already have seen the first two of these definitions in Section 7.1 on entity life histories.

■ An entity is a data object of interest that has been identified in the entity relationship model. In the Just a Line system, customers, orders and cards are all objects of interest, giving us the entities Customer, Order and Product.

■ An event is something that happens in the real world, which may cause an entity (or more than one entity) to be updated. For example, a customer may move house or the price of a product may change. Both of these are events that will, respectively, affect the entities Customer and Product and may cause them to change state.

■ The state of an entity represents a period of time during which the entity satisfies some condition or waits for an event. There are two special states shown in a state diagram: the start state and the stop state. Each state diagram must have only one start state, since all instances of the same entity must begin life in the same state. However, the diagram can have multiple stop states, since the way in which an entity ends its life depends on the sequence of events that it undergoes during its lifetime.

■ A *state transition* represents the response of an entity to an event; the response may involve movement of the entity from one state to another, or the entity remaining in the same state (sometimes referred to as *self-transition*). A state transition is regarded as instantaneous and cannot be interrupted. It consists of three parts: event, guard and action. All of these are optional and will be discussed later in the chapter.

Notation for state diagrams

As an example of a simple state diagram, imagine two balloons, one red and one blue. Both balloons have been blown up, but the blue one has had some of the air let out of it. If someone now tries to blow up the balloons, they will respond in different ways: the blue one will simply become slightly more inflated, but the red one (which is already fully blown up) will burst. We can illustrate these alternative behaviours in a very simple state diagram for the entity Balloon, as shown in Figure 7.11. The figure illustrates the principal components of a state diagram:

■ states – shown by rectangles with rounded corners
■ start state – represented by a large dot
■ stop state – represented by a bull's eye
■ transitions – shown by labelled arrows.

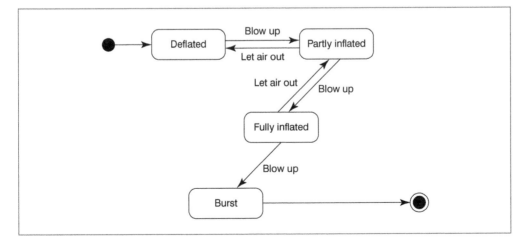

Figure 7.11 Simple state diagram for the entity Balloon

Example state diagrams from Just a Line – entities Customer and Order

Figures 7.12 and 7.13 show simple state diagrams for the Just a Line entities Customer and Order. You can find the equivalent entity life histories for these entities in Figures 7.2 and 7.10 in this chapter.

More advanced state diagrams

More complex information can be shown in a state diagram by using the three optional parts of the state transitions: events, guards and actions.

- An event is something that happens in the real world, which causes an entity to be updated and may cause it to change state.

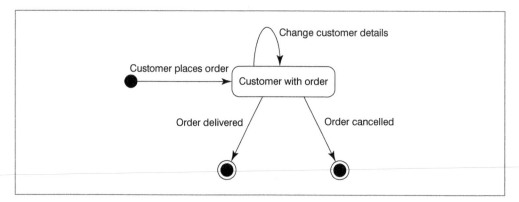

Figure 7.12 State diagram for the Just a Line Customer entity

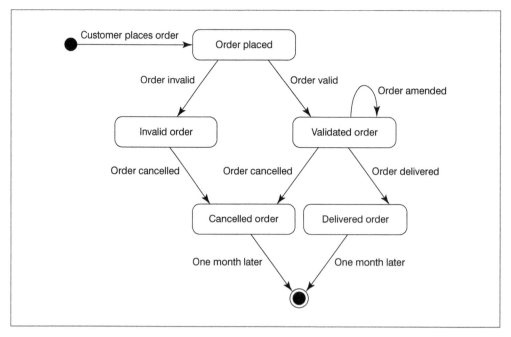

Figure 7.13 State diagram for the Just a Line Order entity

- A guard is a condition that allows the transition to take place only if the condition is true; guards are written inside square brackets [ ].
- An action is behaviour that occurs when a transition takes place, and is preceded by a slash /.

Events can be seen on the transitions in Figures 7.11, 7.12 and 7.13. An example of a guard can be seen in Figure 7.14, which shows a simple state diagram for the Just a Line Product entity. A product is created when it is adopted by Just a Line; this is an event that moves the product into the state of being available. In the available state, the product may be ordered by customers; if the order is for fewer cards than the current number in stock, the product will remain in the available state; however, if the order equals or exceeds the number of cards currently in stock, the product will move to the out-of-stock state. It will remain in this state until it is restocked from the supplier, when it moves back into the available state. The entity Product is deleted when Just a Line stops selling it; this event can occur only when the product is out of stock. Guards can be seen on two transitions in this diagram: Product ordered [order quantity < no. in stock] and Product ordered [order quantity >= no. in stock].

We can add more information to the state diagram by adding an action to the Product state diagram. This is shown in Figure 7.15 where the action /Reorder product is added to the state transition Product ordered [order quantity >= no. in stock].

To finish, we will add one more refinement to this diagram. At any time during the life of a product the price or the number of cards in a packet may change. It would clutter up the diagram if we had to add these two events to both the available and the out-of-stock states, so we enclose both states in a superstate and show the two transitions on that state. Use of the superstate is shown in Figure 7.16.

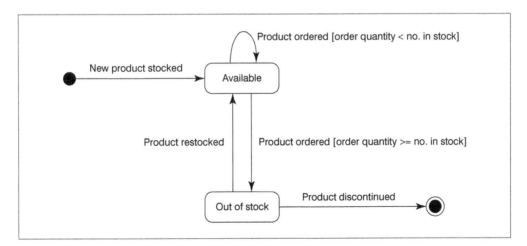

Figure 7.14 Simple state diagram for the Just a Line Product entity, showing guards

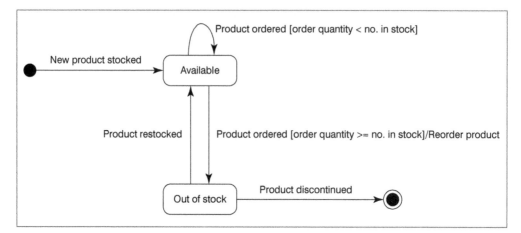

Figure 7.15 State diagram for the Just a Line Product entity, showing guards and action

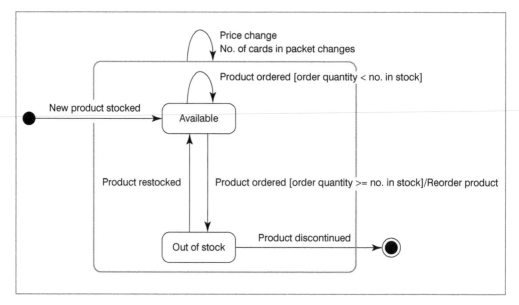

Figure 7.16 State diagram for the Just a Line Product entity, showing use of a superstate

Summary

Entity life histories and state diagrams are techniques that model how the entities in a system respond to external events. Each entity life history or state diagram records all the events that can affect an occurrence of one entity during its lifetime. The techniques record the relative timing of events – for example, an order must be placed before it can be amended or delivered. This information about timing constraints is vital for our understanding of the system, and is not recorded in any of the other system models. The diagrammatic presentation of the information in ELHs and state diagrams makes it easier for the system developer to understand and to explain to the client. The notation for both techniques is simple and straightforward, but the different symbols can be combined to produce complex and expressive models of the developing system.

Exercises and topics for discussion
What can you remember?

You will find all the answers in the chapter

a) What views of the system are provided by (i) data flow diagrams and (ii) the data model?

b) What is an entity?

c) What is an event?

d) What does an asterisk (∗) mean in an entity life history?

e) How is selection represented in an entity life history?

f) What are the three types of effect that an event can have on an entity?

g) What is the role of status indicators in an ELH?

h) What is meant by the state of an entity as shown in a state diagram?

i) What is a state transition?

j) How many start and stop states can you have in a state diagram?

k) What is a guard on a state transition?

l) What is an action on a state transition?

m) When would you use a superstate in a state transition diagram?

7.1 Study the diagram of the Order entity life history (Figure 7.10) and answer the following questions that relate to it.

    (a) Which event creates an occurrence of the Order entity?

    (b) Can an order be cancelled just after it has been placed?

    (c) How many times can an order be amended?

    (d) Is it possible for an order to be delivered and then cancelled?

    (e) If an order is to be deleted, it must be in one of two different states. What are these states?

7.2 The information to be recorded about an entity X is as follows: either event 1 will occur, or event 2 will occur followed by event 3. Which of the diagrams in Figure 7.17, (a) or (b), represents this information correctly?

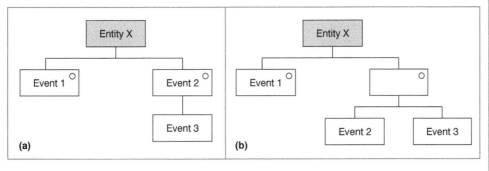

(a) (b)

Figure 7.17 ELH diagrams

7.3 There is a deliberate mistake in the entity life history in Figure 7.18. What is it?

7.4 In the required system for Just a Line, details are recorded about regular customers. An occurrence of the Customer entity is created when a customer places his or her first order. The entity is amended when the customer is added to the mailing list. A customer's details may change during the life of the entity, and finally the entity occurrence is deleted when the customer moves out of the area or when he or she has not placed any order with Just a Line for over a year. Draw an entity life history for the Customer entity in the required Just a Line system.

7.5 In Figure 7.19 you will find an entity life history of an entity Invoice. Describe this life history in English.

7.6 Draw a simple state digram for a Candle entity. A candle is created when it is made; at that stage it is in an unlit state. The candle may be lit, in which case it moves to the lit state, and in the lit state it may be blown out and revert to the unlit state. The entity is deleted when the candle is burnt out.

![Entity X ELH diagram]

Figure 7.18 ELH diagram

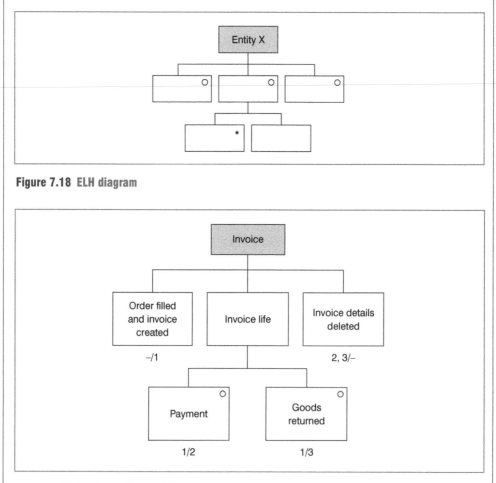

Figure 7.19 Invoice entity life history

7.7 Study the state diagram in Figure 7.20, which illustrates a petrol tank, and then answer the questions that follow.

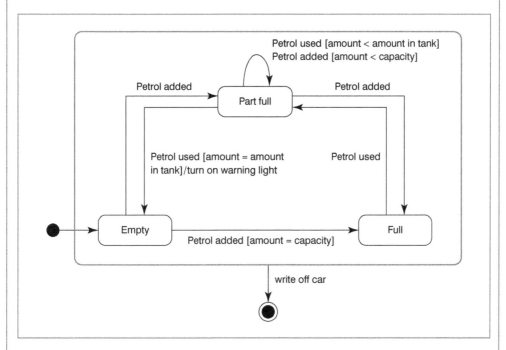

Figure 7.20 State diagram for a petrol tank

> **(a)** What happens if the tank is empty, and then as much petrol is added as it can take?
> **(b)** What happens if the tank is part full and some petrol is used, but there is still some left in the tank?
> **(c)** When is the warning light turned on?
> **(d)** What events can happen when the petrol tank is empty?

7.8 Modify the state diagram for the entity Balloon, shown in Figure 7.11, to include a self-transition on the state Partly inflated, guards on some of the transitions, and a new state, Tied up.

7.9 Study the state diagram in Figure 7.21, which illustrates a machine for selling crisps. For simplicity, this diagram does not include a stop state. Write a brief description of the behaviour of the machine.

7.10 Draw a state diagram to illustrate the behaviour of a child's bank account, where no overdraft is allowed. The account is empty to start with. Money can then be deposited, to put the account in credit, and taken out, as long as the account does not become overdrawn. The account can be closed only when the balance is zero.

7.11 Draw a state diagram to illustrate the behaviour of a kitchen timer, as described below.

To start with, the timer is off. It can be set to move it into the Set state, and turned off to move it back to the Off state; in the Set state the time can be changed. Once the timer is set and the set time reached, the alarm rings. It can be turned off, or will stop automatically after one minute. The timer may break at any time.

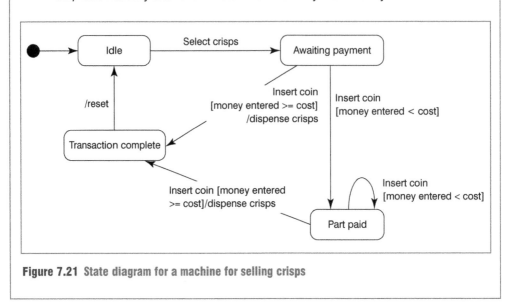

Figure 7.21 State diagram for a machine for selling crisps

References and further reading

Bennett, S., McRobb, S. and Farmer, R. (2005) *Object-Oriented Systems Analysis and Design using UML*, 3rd edn, McGraw-Hill, London.

Goodland, M. and Slater, C. (1995) *SSADM Version 4: A Practical Approach*, McGraw-Hill, London.

Sommerville, I. (2004) *Software Engineering*, 7th edn, Addison-Wesley, Wokingham.

CHAPTER 8
IMPLEMENTATION IN A RELATIONAL DATABASE

In this chapter we look at:

- issues relating to choice of implementation

- a typical commercial database package

- moving from analysis and design models to implementation

- SQL (Structured Query Language)

- types of user interface

- designing a website

After studying this chapter and working through the exercises you will be able to:

- make an informed choice of implementation for a system

- describe the main components of a commercial database package

- explain how models from analysis and design are implemented in a database package

- use SQL to query and manipulate data in a database

- compare and evaluate different types of user interface

- recognise features of an effective website

Introduction

During the software development process there are many issues that must be considered by the developer and many decisions to be taken. Some of the factors that influence these decisions are known from the start of development, others become apparent only as work on the system progresses. Every system is different, with its own individual set of circumstances. In this chapter we introduce some of the issues that a system developer should bear in mind while implementing a system such as Just a Line using a relational database package. We look at Microsoft Access™, a widely used database package, and at Structured Query Language (SQL), which allows users to manage data, get the answers to questions from the database and output the results. Finally, we consider some of the issues that developers have to bear in mind when designing the *user interface* and website for a system such as Just a Line. The chapter is divided into the following sections:

- Types of implementation
- Database
- Microsoft Access: a typical relational database
- Implementing simple entity-relationship diagrams in a database
- SQL (Structured Query Language)
- The user interface
- The Internet.

8.1 Types of implementation

In the early days of systems development almost the only choice of implementation was in a *third-generation language* such as COBOL or FORTRAN. All software had to be written from scratch, which meant that coding took up a large proportion of the overall development time. The data in a system was stored in a number of separate files, which led to huge problems with updating and consistency.

Modern programming languages and databases improved the situation in several ways.

- Today's programming languages are very much more powerful than the languages of previous generations. On average, one line of code today corresponds to ten lines of code in one of the old third-generation languages.
- Older languages are procedural; the programmer has to describe in detail how every task is to be carried out by computer. With some modern languages, however, there is no need to do this; the programmer merely has to define what must be done. In other words, many of today's languages are much more problem oriented than their predecessors.
- Modern languages provide support for *prototyping* (see Chapter 11).
- Modern databases are able to store large amounts of data that can be used in a variety of ways by different applications.

Just a Line is a typical small business system; it involves the storage and manipulation of data about cards, customers, orders and stock, and the production of user-friendly outputs from that data. Although it would be possible to program a system for the company from scratch, this would probably not be a cost-effective proposition. It is much more likely that a developer would use a commercial, widely available database package, such as Microsoft Access. In making this choice the developer must consider all the relevant circumstances. Each system is individual, but it is still useful to weigh up the overall advantages and disadvantages of the different implementation methods.

Advantages of implementing a system such as Just a Line in a programming language

- The client gets a tailor-made system that will cater specifically for his or her particular needs.
- Requirements that are specific to the client organization can be built in to the system.
- The software can be simpler and more compact than software developed with a commercial package, because it is purpose built for a specific system.
- With good documentation, it should be relatively easy to make modifications to the system in the future.

Disadvantages of programming a system for Just a Line

- Effort – if a system is to be programmed, a team of programmers will be given a specification from which they design, code and test the system. This is a very labour-intensive process compared with other methods of development.
- Cost – because of the amount of work involved, a programmed system will nearly always be considerably more expensive than one implemented using a package.
- Time – writing programs and documentation is very time-consuming and can easily cause a development project to over-run predicted deadlines.

Advantages of using a commercial database package for Just a Line

- Packages are a relatively inexpensive method of implementation.
- Using a package saves hours of time and effort that would have to be spent on coding.
- Packages are designed to be *portable*. This means that it is possible for the client to purchase new hardware, yet still run the same software on it.
- Commercial packages have been tried and tested and can usually be guaranteed to work.
- Commercial packages can be installed and run with a minimum of effort. Time is saved on development and documentation.

Disadvantages of using a commercial package

- Although a commercial package supports certain classes of application, some development is still required.
- Commercial packages are limited in the type of application that they will support (although a package would certainly be appropriate for Just a Line).
- Commercial packages are not written for a particular client. It is therefore unlikely that any package will exactly fit a company's needs. Usually, in order to appeal to as many clients as possible, the package will have far more functionality than is required by any one company. The purchaser, therefore, will be paying for functionality he or she does not use.
- The new system has to fit in with what the package can do; where there is a conflict, clients may have to alter their requirements for the system.
- With a package, errors in the system can be difficult to trace and correct.

- Commercial packages are constantly being updated; clients may eventually find that their system is no longer compatible with the latest version of the package software.

Obviously, most of the disadvantages listed above for implementing with a commercial package will apply to Just a Line, but a database package would fit Harry and Sue's present requirements very well and is certainly a cost-effective proposition.

8.2 Database

In older systems, data used to be held in a separate file for each application in the system; in the case of Just a Line, for example, there would have been a file for customers, one for cards, and one for orders. The problem with this is that some items of data are held in more than one place – for example, some of the customer details would have to be stored on both the customer and the order files to link a customer to an order. When a data item is stored in more than one place this is called **data redundancy** (see Chapter 6). Data redundancy should be avoided, first because it is an inefficient use of space and, more importantly, because when data is updated, this has to be done in each place it is stored or the data in the system will be inconsistent. Data today is generally held in a centralized **database**, which is organized in such a way as to facilitate the storage, retrieval and manipulation of data by a number of different applications and users. The way that the data is defined and stored is separated from the various ways in which it is used; this separation of data storage from its applications is called **application independence**.

A database is managed by a **database management system (DBMS)**, which organizes the inputting, manipulation, retrieval and outputting of the data. In the Just a Line system, for example, the DBMS will allow users to enter details of orders (by means of user-friendly **forms**), to find answers to **queries**, such as which cards are ordered most frequently, to perform calculations, such as the total cost of an order, and to produce user-friendly outputs, such as invoices. The DBMS is the link between the way in which the data is stored and the way in which it is used by applications. This is illustrated in Figure 8.1.

There are different types of database, which reflect different ways in which the data is organized, but most databases today are based on a relational model (for an explanation of this, see Begg and Connolly, 2004; Date, 2003; Howe, 2001). A typical commercial database package will include tools that allow the user to:

- query the stored data; most databases use a specialized query language, such as SQL (see Section 8.5)
- create forms for data input
- perform calculations and analyses
- produce user-friendly **reports**.

In the next section, we look in more detail at a popular commercial database package that would be appropriate for the Just a Line system development.

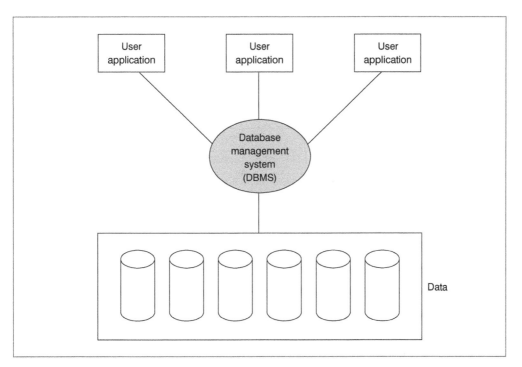

Figure 8.1 The DBMS as link between applications and data

8.3 Microsoft Access: a typical relational database

We briefly describe below some of the features of Microsoft Access, a popular relational database.

Tables

A database in Microsoft Access is built around **tables**; all other parts of the database, such as queries, forms and reports, depend on the data in the tables. Tables store data in a row–column format; each column stores a field, or attribute of the data, and each row stores a record, typically the complete set of values for a single data object. Figure 8.2 shows a small example table that contains data about customers.

Every table should have a primary key (an attribute, or combination of attributes, that uniquely identifies each record). In Figure 8.2, the primary key is Customer Number.

Queries

The tables in the database contain all the data that the users need, but not in a very accessible form. Users frequently want to view data structured in a particular way, or they want to retrieve only those records that satisfy a certain condition. In order to allow this, Access provides the query mechanism, which enables users to retrieve and analyse the data in an appropriate and convenient way. Figure 8.3 illustrates the result of a query on the data in Figure 8.2, requesting details of customers who live in Hansford.

Customer						
CustomerNo	**Name**	**FirstName**	**Street**	**Town**	**Postcode**	**PhoneNumber**
1	Leary	John	14 High St.	Hansford	SG7 4DG	01483 876594
2	Trip	Chris	2 Long Ave.	Boxeth	SG8 5TR	01485 324906
3	Jones	Ellen	67 Bow Rd.	Hansford	SG7 4RF	01483 897885
4	Brown	Lisa	124 High St.	Hansford	SG7 4DF	01483 874553
5	James	Bob	21 Park Rd.	Boxeth	SG8 7YH	01485 347236

Figure 8.2 Example of a small table of customers in an Access database

HansfordCustomer						
CustomerNo	**Name**	**FirstName**	**Street**	**Town**	**Postcode**	**PhoneNumber**
1	Leary	John	14 High St.	Hansford	SG7 4DG	01483 876594
2	Jones	Ellen	67 Bow Rd.	Hansford	SG7 4RF	01483 897885
3	Brown	Lisa	124 High St.	Hansford	SG7 4DF	01483 874553

Figure 8.3 Result of a query requesting details of customers who live in Hansford

Forms

Tables are the core of the database, but they are not very user-friendly when it comes to viewing and manipulating data. A form in Access allows the user to enter, modify and display data on screen. Forms are based on tables, but may include labels, boxes, lines and even pictures to make the user's task easier. Figure 8.4 shows a form, based on the table in Figure 8.2, which displays the first record in the table.

Reports

Reports allow the user to format and display information that is output from the system in a user-friendly form. Reports are like forms in that they may include labels, boxes, lines and pictures. Figure 8.5 shows a completed report invoice for a customer in the Just a Line system.

Macros

It is often useful to automate tasks, such as printing a report, that are performed on a regular basis. A macro is a series of actions that are used to carry out a task and that have been pre-recorded, so that the user has only to press a command button.

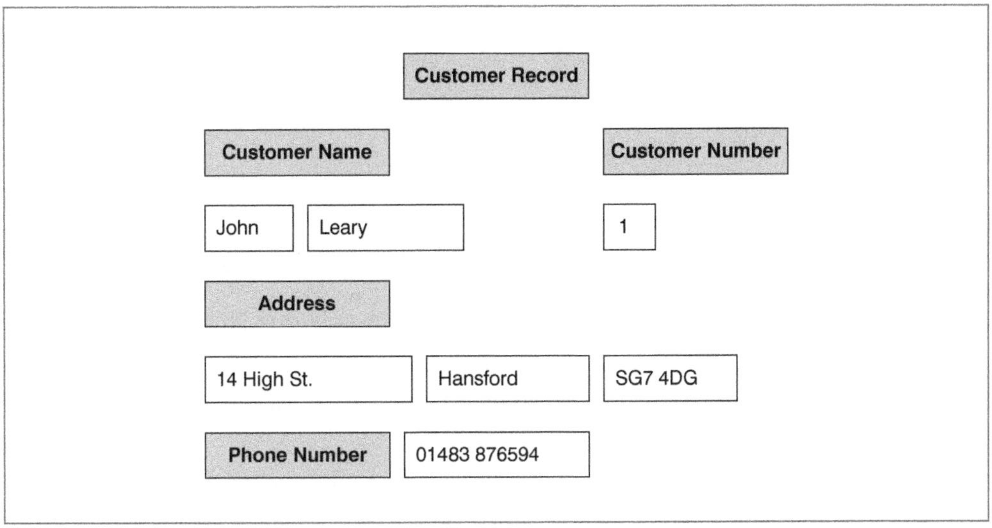

Customer Record

Customer Name

John	Leary

Customer Number

1

Address

14 High St.	Hansford	SG7 4DG

Phone Number	01483 876594

Figure 8.4 Form displaying the first record from the table in Figure 8.2

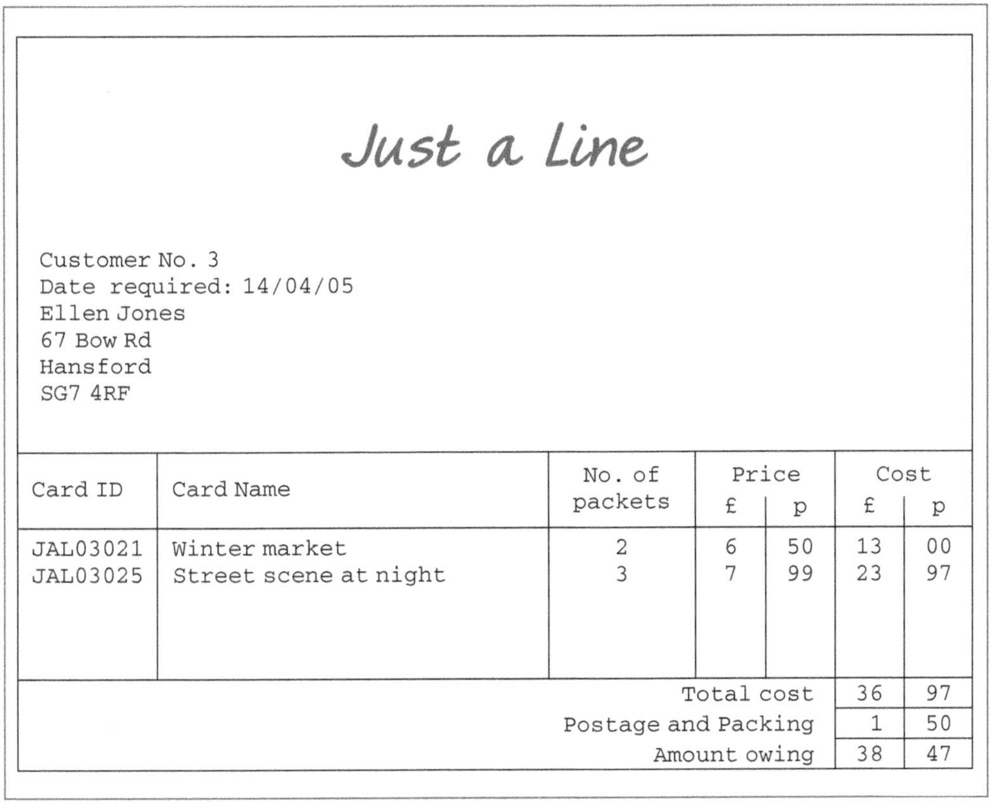

Just a Line

Customer No. 3
Date required: 14/04/05
Ellen Jones
67 Bow Rd
Hansford
SG7 4RF

Card ID	Card Name	No. of packets	Price £	Price p	Cost £	Cost p
JAL03021	Winter market	2	6	50	13	00
JAL03025	Street scene at night	3	7	99	23	97
		Total cost			36	97
		Postage and Packing			1	50
		Amount owing			38	47

Figure 8.5 A report invoice for the Just a Line system

Modules

A module is a means of organizing code written in Visual Basic (VB), a programming language developed by Microsoft and based on the BASIC language. It is a collection of statements and procedures that can perform a specific task and that are stored together as a single, named unit.

Data access pages

A data access page is a web page that has a connection to an Access database and allows the user to view and manipulate the data in the database via the Internet.

8.4 Implementing simple entity-relationship diagrams in a database

It is beyond the scope of this book to describe a full implementation of the Just a Line system in a database package such as Microsoft Access. In this section we briefly cover some of the main issues involved. You can find more information about designing and implementing a database information system in the books listed in the further reading section at the end of this chapter.

The principal source of information for the developer when deciding how the required data is to be stored in tables is the entity-relationship diagram (see Chapter 6). The developer will start by constructing a table for each entity in the diagram; so, for example, tables in the Just a Line system will include **Customer**, **Product** and **Order**. In general the attribute that was identified as the primary key of the entity becomes the primary key of the table, and the other entity attributes become fields. Figure 8.6 shows the **Product** entity as identified in Chapter 6, Figure 8.7 shows the **Product** table in Access in design view and Figure 8.8 shows it in datasheet view. As Just a Line currently uses only one supplier, all the entries in the last column of the table are the same, but this design allows details of new suppliers to be added if required.

Once the entity tables have been constructed, the developer has to decide how the relationships between them are to be implemented. In a one-to-many relationship this may be achieved by including the primary key from the table of one of the entities in the relationship as a foreign key in the table of the other entity. Figure 8.9 shows the one-to-many **Supplier to Product** relationship from the entity-relationship diagram.

If you now look again at Figure 8.7, which shows the **Product** table in design view, you can see that the primary key of **Supplier** (**SupplierNo**) has been included. It acts as a foreign key in the table to implement the **Supplier to Product** relationship.

Product

productCode
productDescription
packetSize
costPerPacket
supplier#

Figure 8.6 The Product entity as identified in Chapter 6

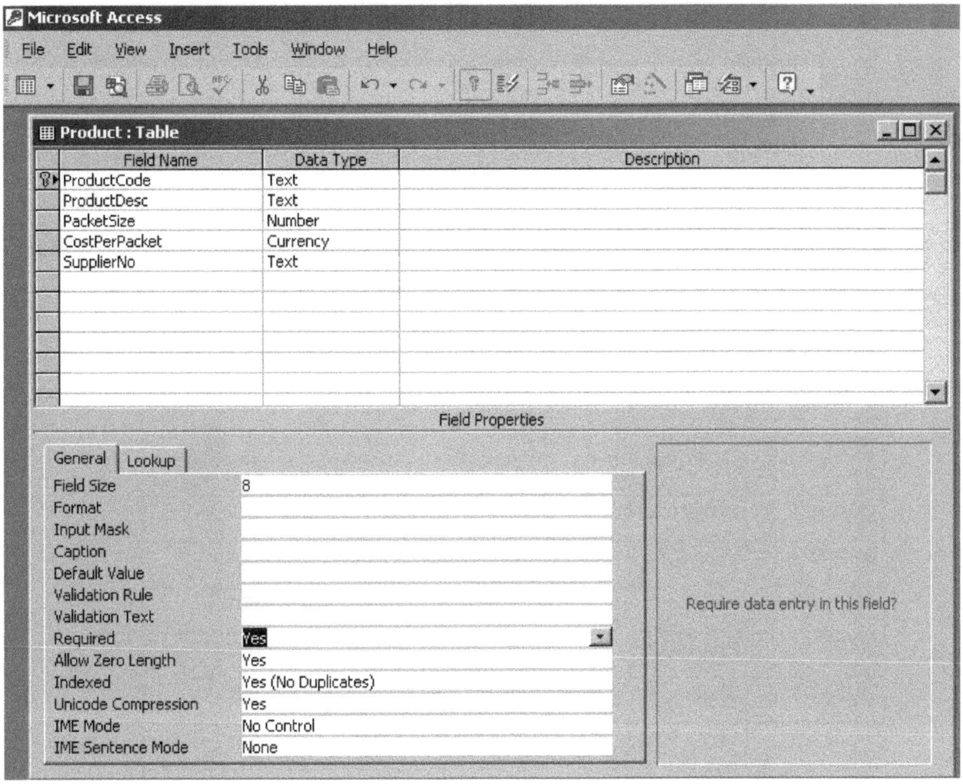

Microsoft Access

File   Edit   View   Insert   Tools   Window   Help

Product : Table

Field Name	Data Type	Description
ProductCode	Text	
ProductDesc	Text	
PacketSize	Number	
CostPerPacket	Currency	
SupplierNo	Text	

Field Properties

General | Lookup

Field Size	8
Format	
Input Mask	
Caption	
Default Value	
Validation Rule	
Validation Text	
Required	Yes
Allow Zero Length	Yes
Indexed	Yes (No Duplicates)
Unicode Compression	Yes
IME Mode	No Control
IME Sentence Mode	None

Require data entry in this field?

Figure 8.7 The Access Product table in design view

Product				
ProductCode	ProductDesc	PacketSize	CostPerPacket	SupplierNo
JAL03021	Winter market	5	6.50	S01
JAL03022	Fairground	10	6.50	S01
JAL03023	Carnival	10	7.99	S01
JAL03024	Procession	10	5.49	S01
JAL03025	Street scene at night	5	7.99	S01
JAL03026	By the waterfront	10	9.49	S01

Figure 8.8 The Product table in datasheet view

Figure 8.9 The one-to-many, Supplier to Product relationship from the entity-relationship diagram

149

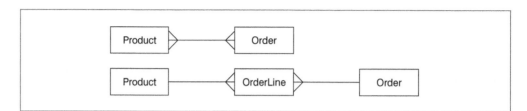

Figure 8.10 The Product to Order, many-to-many relationship split into two one to many relationships

Product				
ProductCode	**ProductDesc**	**PacketSize**	**CostPerPacket**	**SupplierNo**
JAL03021	Winter market	5	6.50	S01
JAL03022	Fairground	10	6.50	S01
JAL03023	Carnival	10	7.99	S01
JAL03024	Procession	10	5.49	S01
JAL03025	Street scene at night	5	7.99	S01
JAL03026	By the waterfront	10	9.49	S01

Order		
OrderNo	**CustomerNo**	**DateRequired**
051	3	12/05/2005
052	5	16/05/2005
053	1	16/05/2005

OrderLine		
OrderNo	**ProductCode**	**NoOfPackets**
051	JAL03021	10
051	JAL03024	2
051	JAL03025	5
052	JAL03024	10
053	JAL03025	5
053	JAL03026	1

Figure 8.11 Tables Product, Order and OrderLine

In the case of a many-to-many relationship, it is necessary to construct a separate table to implement the relationship in the database. It is likely that this process will have been carried out at the stage of constructing the entity-relationship diagram. There is an example of this in Chapter 6, where the many-to-many relationship **Product to Order** is split into two one-to-many relationships – **Product to OrderLine** and **Order to OrderLine** (see also Figure 8.10).

If the many-to-many relationships have not been removed from the entity-relationship diagram, this must be done at the database design stage. The resulting tables **Product**, **Order** and **OrderLine** are shown in Figure 8.11. The primary key of the table **OrderLine** is made up of the primary keys of the other two tables.

Of course, the storage of data is only the start of the implementation. It is equally important for the developer to ensure that the final system can support all the functionality that is required. In a database system such as Microsoft Access, the functionality is implemented by means of queries, which retrieve and manipulate data from the tables.

The developer's sources for implementing the functionality of the system are the data flow diagrams (mainly the lower-level processes) and the process definitions (see Chapter 4).

As an example, Figure 8.12 shows a level 2 data flow diagram containing the process **Price order**. In order to implement the **Price order** process for a particular order, the system will first have to calculate the cost of each separate line on the order.

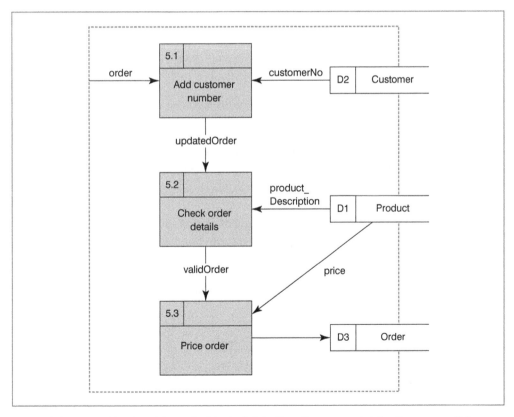

Figure 8.12 Just a Line Level 2 required logical data flow diagram containing the process Price order; the original flow productDetails has been decomposed into productDescription and price, which would be documented in the data dictionary

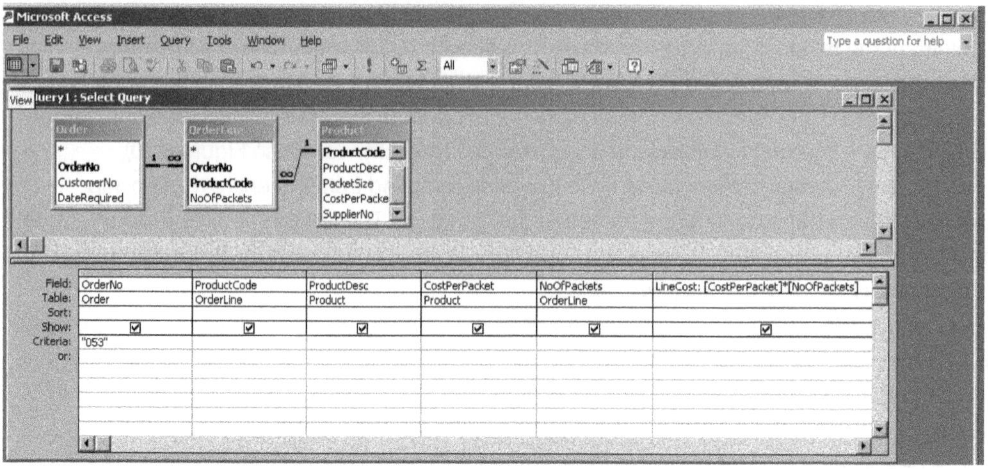

Figure 8.13 Query to price the order lines for order number 053

LineCost					
OrderNo	**ProductCode**	**ProductDesc**	**CostPerPacket**	**NoOfPackets**	**LineCost**
053	JAL03025	Street scene at night	7.99	5	39.95
053	JAL03026	By the waterfront	9.49	1	9.49

Figure 8.14 Table showing the result of a query to price the order lines in order number 053

Using Access, if we want to find out the cost of each order line for order number 053, we need to set up a query that takes fields from the **Product**, **Order** and **OrderLine** tables as shown in Figure 8.13; the calculation of the cost of the line **(CostPerPacket* NoOfPackets)** is in the final column. This query will produce a new table showing the cost of each line in the order (see Figure 8.14).

A more sophisticated query will also total the costs of each product ordered, add the delivery charge or postage, and calculate the total amount owed by the customer on this order.

In the following section we look at SQL, the standard database language, which is part of almost all databases and provides the functionality that users require.

8.5 SQL (Structured Query Language)

When working with a database users need to be able to:

- create the database and its internal structures
- manage the data by performing operations such as insert, update, delete and sort
- query the data and output the results.

SQL (Structured Query Language) is the standard database language, which allows users to do all of the tasks listed above. SQL is portable, so it can be used with a range of database management systems. It is also relatively user-friendly and easy to learn, so it can be used by a range of people, from database administrators to end-users.

SQL has two parts: a DDL (data definition language) that allows users to define the database structure, and a DML (data manipulation language) that allows users to retrieve and manipulate data. In this section we focus on the DML and the commands that allow users to work with the data in the database.

In the previous section we looked at how the **Product** table structure can be created in Microsoft Access (see Figure 8.7). If you want to see how tables are created in SQL, you can find more information in Begg and Connolly (2004).

SQL has the following commands for data manipulation, which allow users to manage and query the data in the database:

```
INSERT — to add data to a table
DELETE — to remove data from a table
UPDATE — to modify the data in a table
SELECT — to query the data and produce the result
```

Insert

In order to populate a database table with data, we use the INSERT command, as shown in the example below. Data items that are enclosed in single quotes, such as **'Winter Market'**, are character literals.

Note that all SQL commands must end with a semi-colon ';'.

```
INSERT INTO     Product

VALUES          ('JAL03021','Winter Market',5,6.50,'S01');
```

You can see the completed Product table in Figure 8.8.

Update

The UPDATE command allows us to change entries in the table. The following example shows how to increase the cost of all packets in the **Product** table by 5 per cent.

```
UPDATE     Product
SET        CostPerPacket = CostPerPacket*1.05;
```

If we want to change only some of the entries – for example, those where the packet size is 10 cards – we add a WHERE clause, as shown below.

```
UPDATE     Product
SET        CostPerPacket = CostPerPacket*1.05
WHERE      PacketSize = 10;
```

The **Product** table will now look as shown in Figure 8.15.

Product				
ProductCode	ProductDesc	PacketSize	CostPerPacket	SupplierNo
JAL03021	Winter market	5	6.50	S01
JAL03022	Fairground	10	6.83	S01
JAL03023	Carnival	10	8.39	S01
JAL03024	Procession	10	5.77	S01
JAL03025	Street scene at night	5	7.99	S01
JAL03026	By the waterfront	10	9.96	S01

Figure 8.15 The updated Product table

Delete

The DELETE command allows us to remove entries from the table, as in the following example, which removes the record for the product with code JAL03023.

```
DELETE FROM      Product
WHERE            ProductCode = 'JAL03023';
```

If we wish to remove all the records, but still preserve the table structure, we omit the WHERE clause, as shown below:

```
DELETE FROM      Product;
```

Select

The most powerful of the commands in SQL's data manipulation language is SELECT, which allows us to query the database, retrieve and display the data. The result of a SELECT query on a table is always another table, which allows users to carry out further queries.

We do not have room here to describe everything that SELECT can do, but you will be able to get some idea from the examples below; the first three of these are based on the Customer table from the Just a Line system. You can see this in Figure 8.2 and it is repeated below in Figure 8.16.

To display the whole Customer table:

```
SELECT *
FROM Customer;
```

To display the first name, name and phone numbers of all customers in the **Customer** table:

```
SELECT FirstName, Name, PhoneNo
FROM Customer;
```

The result of this query is shown in Figure 8.17.

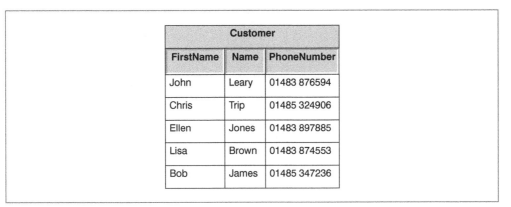

Customer						
CustomerNo	Name	FirstName	Street	Town	Postcode	PhoneNumber
1	Leary	John	14 High St.	Hansford	SG7 4DG	01483 876594
2	Trip	Chris	2 Long Ave.	Boxeth	SG8 5TR	01485 324906
3	Jones	Ellen	67 Bow Rd.	Hansford	SG7 4RF	01483 897885
4	Brown	Lisa	124 High St.	Hansford	SG7 4DF	01483 874553
5	James	Bob	21 Park Rd.	Boxeth	SG8 7YH	01485 347236

Figure 8.16 The Customer table from the Just a Line system

Customer		
FirstName	Name	PhoneNumber
John	Leary	01483 876594
Chris	Trip	01485 324906
Ellen	Jones	01483 897885
Lisa	Brown	01483 874553
Bob	James	01485 347236

Figure 8.17 Result of a query requesting first name, name and phone numbers of all customers in the Customer table

To select complete records for only those customers who live in Hansford:

```
SELECT *
FROM Customer
WHERE Town = 'Hansford';
```

You can see the result of this query in Figure 8.18.

If we want to display the different towns that customers come from, we can use the command:

```
SELECT Town
FROM Customer;
```

However this will display all the occurrences of each town, including duplicates, as shown in Figure 8.19.

HansfordCustomer						
CustomerNo	Name	FirstName	Street	Town	Postcode	PhoneNumber
1	Leary	John	14 High St.	Hansford	SG7 4DG	01483 876594
2	Jones	Ellen	67 Bow Rd.	Hansford	SG7 4RF	01483 897885
3	Brown	Lisa	124 High St.	Hansford	SG7 4DF	01483 874553

Figure 8.18 Result of a query requesting details of customers who live in Hansford

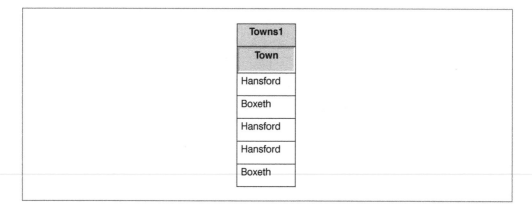

Towns1
Town
Hansford
Boxeth
Hansford
Hansford
Boxeth

Figure 8.19 Result of a query displaying all the occurrences of each town, including duplicates

Towns2
Town
Boxeth
Hansford

Figure 8.20 Result of a query displaying each town, without duplicates

In order to display each different town once only, we use the DISTINCT keyword as below. The resulting table is shown in Figure 8.20.

```
SELECT DISTINCT Town
FROM Customer;
```

The SELECT command can be used to produce calculated fields. We will use the **Product** table (see Figure 8.21) to give an example of this.

Product				
ProductCode	**ProductDesc**	**PacketSize**	**CostPerPacket**	**SupplierNo**
JAL03021	Winter market	5	6.50	S01
JAL03022	Fairground	10	6.50	S01
JAL03023	Carnival	10	7.99	S01
JAL03024	Procession	10	5.49	S01
JAL03025	Street scene at night	5	7.99	S01
JAL03026	By the waterfront	10	9.49	S01

Figure 8.21 The Product table

The command below will calculate the VAT on the cost of each packet in the table and display this in a new column, as shown in Figure 8.22. The expression **CostPerPacket∗0.175** does the calculation and the expression **AS VAT** displays the result in a new column labelled **VAT**:

```
SELECT ProductCode, ProductDesc, PacketSize, CostPerPacket,
CostPerPacket*0.175 AS VAT, SupplierNo
FROM Product;
```

We can also use SELECT to perform aggregate functions, such as counting records, or finding the total, average, maximum or minimum of a series of values.

The command below gives us the number of records in the **Product** table. The item in brackets after COUNT (**ProductCode** in this example) denotes the field to be counted. As in the previous example, we can use the **AS** keyword to name the new column.

VAT on products					
ProductCode	**ProductDesc**	**PacketSize**	**CostPerPacket**	**VAT**	**SupplierNo**
JAL03021	Winter market	5	6.50	1.14	S01
JAL03022	Fairground	10	6.50	1.14	S01
JAL03023	Carnival	10	7.99	1.40	S01
JAL03024	Procession	10	5.49	0.96	S01
JAL03025	Street scene at night	5	7.99	1.40	S01
JAL03026	By the waterfront	10	9.49	1.66	S01

Figure 8.22 Result of a query to calculate the VAT on the cost of each packet and display this in a new column called VAT

```
SELECT COUNT (ProductCode) AS ProductCount
FROM Product;
```

This will produce the table shown in Figure 8.23.

A further use of the SELECT command is to sort data. If we take the **Product** table as an example (see Figure 8.21), the following command will sort the data about cards in descending order of the cost per packet:

```
SELECT *
FROM Product
ORDER BY CostPerPacket DESC;
```

This will produce the table shown in Figure 8.24.

All of the examples of the SELECT command so far have used only one table, but frequently queries need data from two or more tables in the database. The following example uses the **Product** and **OrderLine** tables (see Figure 8.11) to produce details of products that are currently on order. The expressions **OrderLine.ProductCode** and **Product.ProductCode** distinguish between the **ProductCode** fields in the **OrderLine** and **Product** tables.

The last line of the command **WHERE OrderLine.ProductCode = Product.ProductCode** is the join condition. A join condition always relates to fields that the tables have in common. There must be at least one join condition when using two or more tables.

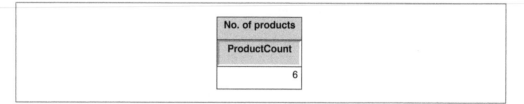

No. of products
ProductCount
6

Figure 8.23 Result of a query to count the number of products in the Product table and display the result in a new column called ProductCount

Sorted product				
ProductCode	ProductDesc	PacketSize	CostPerPacket	SupplierNo
JAL03026	By the waterfront	10	9.49	S01
JAL03025	Street scene at night	5	7.99	S01
JAL03023	Carnival	10	7.99	S01
JAL03022	Fairground	10	6.50	S01
JAL03021	Winter market	5	6.50	S01
JAL03024	Procession	10	5.49	S01

Figure 8.24 The Product table, sorted in descending order according to cost per packet

```
SELECT OrderLine.ProductCode, ProductDesc, CostPerPacket,
   NoOfPackets
FROM OrderLine, Product
WHERE OrderLine.ProductCode = Product.ProductCode;
```

This command will produce the table shown in Figure 8.25.

We will now extend the example above to include the dates that these orders are required. For this query we will have to include the **Order** table, since that is where the dates are stored. The query will have two join conditions because we are using three tables.

```
SELECT OrderLine.ProductCode, ProductDesc, CostPerPacket,
   NoOfpackets, DateRequired
FROM OrderLine, Product, Order
WHERE OrderLine.ProductCode = Product.ProductCode
AND Order.OrderNo = OrderLine.OrderNo;
```

The result of this will be the table shown in Figure 8.26

As a final example, we will use the **Customer**, **Product**, **Order** and **OrderLine** tables to find out the names of customers who have ordered five or more packets of cards and which cards they have ordered. This command has three join conditions (because we are using four tables) and an extra constraint in the expression **NoOfPackets** >= 5. We will display the results in alphabetical order of the customer names. The result of the query is shown in Figure 8.27.

```
SELECT Customer.Name, ProductDesc, NoOfPackets
FROM Order, OrderLine, Product, Customer
WHERE OrderLine.ProductCode = Product.ProductCode
AND OrderLine.OrderNo = Order.OrderNo
AND Customer.CustomerNo = Order.CustomerNo
AND NoOfPackets >= 5
ORDER BY Customer.Name ASC;
```

Products on order			
ProductCode	ProductDesc	CostPerPacket	NoOfPackets
JAL03021	Winter market	6.50	10
JAL03024	Procession	5.49	2
JAL03024	Procession	5.49	5
JAL03025	Street scene at night	7.99	5
JAL03025	Street scene at night	7.99	1
JAL03026	By the waterfront	9.49	10

Figure 8.25 Result of a query on the Product and Order Line tables showing details of products that are currently on order

Orders and dates				
ProductCode	ProductDesc	CostPerPacket	NoOfPackets	DateRequired
JAL03021	Winter market	6.50	10	12/05/2005
JAL03024	Procession	5.49	2	12/05/2005
JAL03025	Street scene at night	7.99	5	12/05/2005
JAL03024	Procession	5.49	10	16/05/2005
JAL03025	Street scene at night	7.99	5	16/05/2005
JAL03026	By the waterfront	9.49	1	16/05/2005

Figure 8.26 Result of a query on the Product, Order and Order Line tables showing details of products that are currently on order and the dates that they are required

Orders of at least 5 packets		
Name	ProductDesc	NoOfPackets
James	Procession	10
Jones	Street scene at night	5
Jones	Winter market	10
Leary	Street scene at night	5

Figure 8.27 Result of a query on the Customer, Product, Order and Order Line tables to find out the names of customers who have ordered five or more packets of cards and which cards they have ordered

From these last few queries you can see that, although SQL is made up of simple components, when these are used in combination it becomes a very powerful database language.

8.6 The user interface

The user interface is the link between the human user and the computer system. It includes screens, reports, documentation, online help facilities and websites – in fact any part of the system with which the user comes into contact.

There are many different types of problem that are found in the interfaces to software systems. Some of the most common are listed below.

■ The displays are cluttered. This may be caused by trying to include too much information, or by inappropriate use of lines, boxes and colour on the screen.

■ The ways in which the user is expected to carry out various tasks may not be consistent.

■ The terminology used for on-screen instructions is confusing.

■ The error messages given by the system are patronizing and unhelpful.

■ The online help and the manual are so convoluted that they might as well be in Ancient Greek, as far as the average user is concerned.

The importance of a good user interface cannot be overemphasized. It does not matter how efficiently a system runs, how reliable it is or how easy it is to maintain – if nobody wants to use it, it is worthless. Today, almost everyone comes into contact with a computer in some way – even if it is only as the recipient of a computerized bill. The days when only trained technicians handled machines are long over, and increasing numbers of people are using computers as an essential part of their work and leisure. Now, more than ever, it is vital that computer systems have the sort of interface that will make people keen to use them, and, in order to achieve this, user interface issues must be considered at all stages of system development.

The user interface and the automation boundary

The scope of the interface for a particular software system corresponds to its automation boundary, which encloses the area of the system that is to be computerized. All the inputs and outputs of the computerized system will cross the automation boundary.

When deciding where the automation boundary lies, it is important to consider the allocation of functions between the computer and the user. For example, in the case of Just a Line, Harry and Sue have decided that they wish to computerize only the customer orders and marketing sections of the business. In Figure 8.28 you can see a level 1 data flow diagram that shows the automation boundary for this reduced system.

A decision to computerize only the customer orders and marketing parts of the Just a Line system means that the task of checking whether cards ordered by customers are actually in stock will have to be performed manually and this information will have to be entered manually into the computer. The user interface of the Just a Line system will have to cater for this – perhaps with a screen such as the one shown in Figure 8.29.

Modifications to the system will frequently alter the automation boundary and involve changes to the user interface. For example, at some time in the future, Harry and Sue may decide that they want to include stock control in the Just a Line system. In this case, checking whether cards that have been ordered are available or not will be done automatically, without involving the user. The interface will simply have to issue a warning message if any cards ordered are not in stock.

The user of the system

As its name implies, the user interface of a system must be built to suit the user. This means that all parts of the system that are visible to the user should reflect the user's view of what is going on and how tasks are carried out. Task analysis is used to help the system developer to gain a sound understanding of the activities currently performed by users, and then to translate these into equivalent tasks in a computerized system. In designing the user interface, the system developer must take into consideration such factors as the amount of previous experience the user has had with computer systems, the user's attitude to the new system and the amount of time he or she will make use of it.

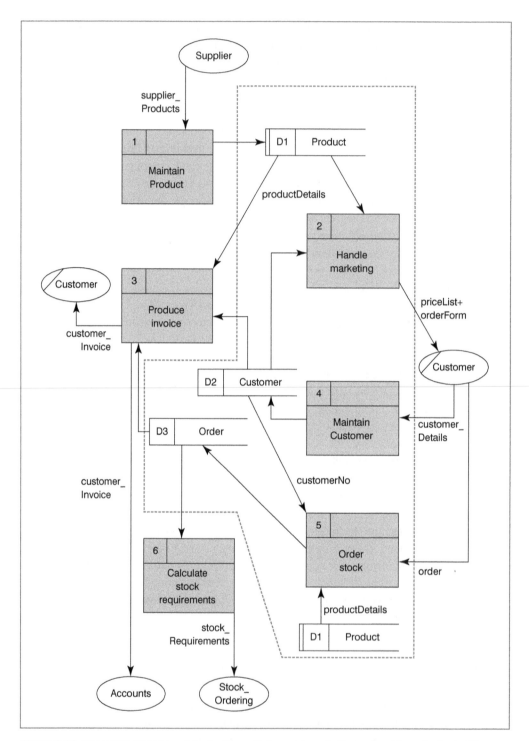

Figure 8.28 The automation boundary for the ordering and marketing sub-systems in Just a Line

Cards in stock			
Card Number	Number of packets in stock	On customer order	On order from supplier
JAL03021	125	25	0
JAL03022	80	30	0
JAL03023	105	65	100
JAL03024	0	10	100
JAL03025	25	30	150
JAL03026	85	40	50
			● Return to main menu

Figure 8.29 A screen to interface with the manual task of checking that cards ordered are in stock

Building the user interface is often a complicated juggling act, since there is nearly always more than one user. The client, who has commissioned the system, may have a managerial role in the organization and may therefore be an infrequent user of the new system. In some cases the developer may find that it is necessary to design a user interface that will support and encourage complete novices, while at the same time satisfying experienced computer professionals. The Just a Line system does not have this problem, since neither Harry nor Sue has had much previous computer experience. However, it would be a mistake to treat them simply as one user. First, they perform different roles in the Just a Line organization. Second, they may be completely different in their ability to learn and use the new system, and in their preferences regarding areas such as screen layout or online help facilities.

Whether there is one user or many, there is little doubt that the most effective way to design the user interface of a system is by prototyping. Diagrams on paper can give some idea of what individual parts of the system will look like, but they are static and so cannot give any idea of how the system will feel to an individual user. The technique of prototyping, which is discussed in Chapter 11, creates a dynamic model that allows the user to have hands-on experience and give feedback about the system before it is built.

Designing screens

One of the areas where prototyping with the user is most effective is the screen design for the new system. The user can see exactly what the system will look like and can express preferences at a stage of development where it is still a simple matter to incorporate changes and new ideas. Working with the user, the developer can make suggestions and ensure that the final screen design not only reflects what the user wants but also that it conforms to the basic guidelines for good screen design. Some of these guidelines are listed below.

■ The screen should be restful to look at, without dramatic colour combinations or a large number of flashing signals. Effects of this kind may be fun in a new system, but very soon become tiresome and distracting.
■ Highlighting should be used only to pick out important information, not to add decoration to the screen.
■ The screen should contain all the relevant information and no more. An overcrowded screen is tiring to look at and irritating to work with.

- The screen should be self-sufficient and self-explanatory. It should not be necessary to refer constantly to either online help or to the user manual to find out what to do next or how to escape out of the screen.
- The overall design of the screens in a system should be consistent – for example, error messages should always appear on the screen in the same position and in the same format.
- If possible, the screen layout should reflect the layout of the associated source documents, for ease of input.
- Data on the screen should be presented in a logical sequence for the user. Related items, such as names and addresses, should be grouped together.
- It should always be obvious how to get out of trouble and how to find more help with particular tasks.

Interaction styles

The *interaction style* of the user interface determines the user dialogue with the system. In the early years of systems development, all interaction between the user and the system was carried out by means of a command language. The user had to type in commands, which were often lengthy and apparently meaningless, to tell the computer what was to be done. Since almost all computer users at that time were trained programmers, they were not troubled by having to type in commands such as **cp Cust.mod RegCust.mod** to copy a file. For a user who is fluent in the particular command language used, the facility of issuing commands to the system gives more control and freedom than are found in other interaction styles.

For those users who are not computer professionals, and particularly for the inexperienced user, a command language interaction style can be intimidating and offputting. Fortunately, in today's software systems, there are alternatives that give the novice user support and encouragement in getting the most out of the system. The most common example of a user-friendly interaction style is *direct manipulation*, where the user actually moves objects on the screen corresponding to the tasks to be carried out. For example, users can copy a file simply by dragging the file icon on the screen across from one disk to another, and delete a file simply by dragging its icon to the on-screen recycle bin. The most widespread and commercially successful use of direct manipulation can be seen in the millions of computer games that are sold each year. Even children barely old enough to read can get hours of enjoyment from making the game hero jump over castle walls, slay monsters and finally find the key to the treasure chest.

For the Just a Line system a command language interaction style would not be suitable, since Harry and Sue have no experience of computer systems. They would probably have a lot of fun with a customized direct manipulation style, but would not be too happy about the time the development would take or how much it would cost. We said in an earlier section that the Just a Line system is very suitable for implementation with a commercial database package. Most packages dictate their own interaction style. This is one that the makers of the package have chosen as suitable for the type of system that the package will be used for. In many packages, the user's interactions with the computer are performed by means of menu selection. This means that the user is presented with a menu of possible options on the screen. When an option is chosen, a further menu of related sub-options may be displayed for the user to make another selection, or the particular task chosen may be carried out.

Menu selection offers a hierarchical route map through the system, which most users find easy to follow. Figure 8.30 shows a sub-set of the menu hierarchy from the Just a Line system.

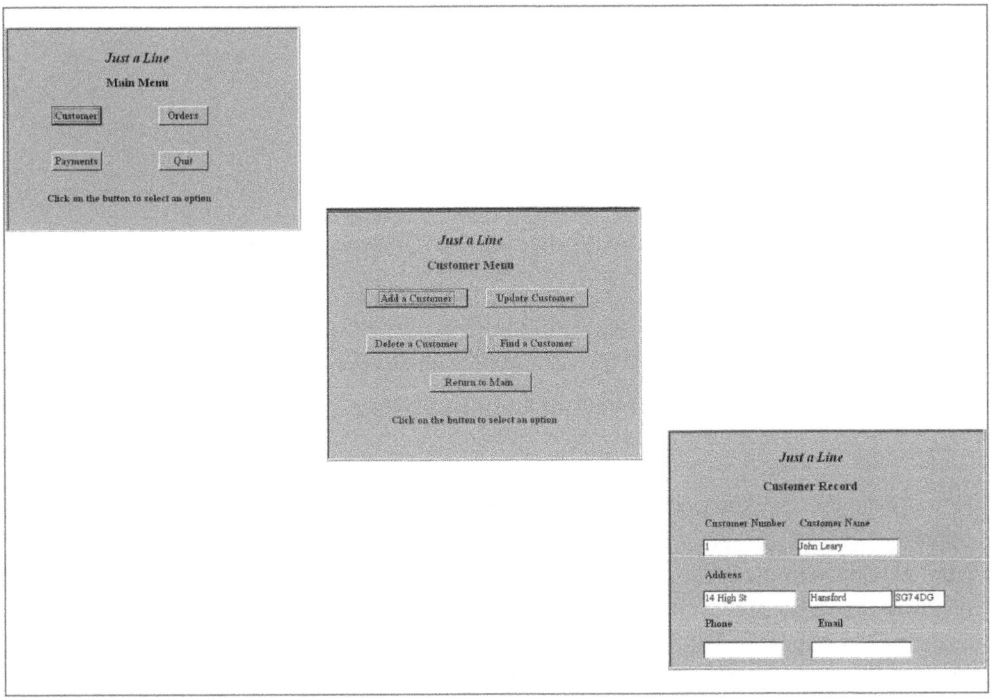

Figure 8.30 Part of the menu hierarchy from the Just a Line system

At the top level, the menu options reflect the main areas of activity of the system: Customers, Orders and Payments; there is also a button to allow the user to leave the system. The user goes to the required option by selecting the appropriate button, such as **Customer**. A screen then appears showing the tasks associated with maintaining customers in the Just a Line system – the user can add a new customer, delete a customer, update a customer's details or find a specific customer record; there is also an option to return to the top-level menu. Selection of the task to be performed is carried out in the same way as the main screen, by clicking on the appropriate button. Selecting **Add a Customer** brings up the customer record screen, which enables the user to enter the customer number and details for the new customer.

Although a commercial package constrains the developer's choice in the design of the user interface for a particular system, there are still many issues to be determined and many decisions to be taken. Some of the factors to be considered in a menu-driven interface are listed below.

■ How many items should appear on a menu?
■ In what order should the options be listed?
■ How should the options be named or described?
■ What sort of prompt should be given to the user?

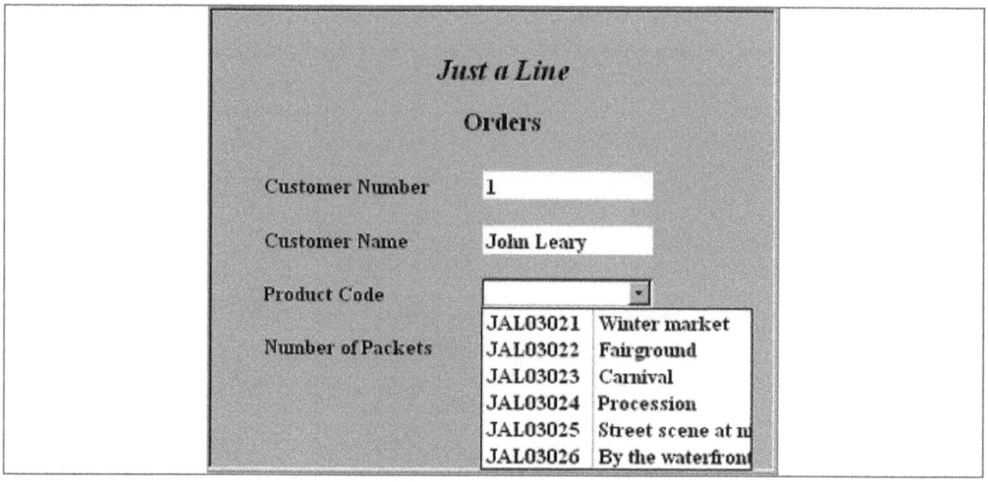

Figure 8.31 Drop-down menu example from the Just a Line system

- How should the user respond: by typing in a number or by highlighting the required option?
- How should the system react if the user does not select a valid option? What sort of error handling is appropriate?
- Does the use of graphics and colour enhance the menu screen or act as a distraction?
- Should a quick route through the menus of the system be provided for experienced users?
- How should help be provided?

Figure 8.31 shows a drop-down menu from the Just a Line system, allowing the user to select from a list showing the products that are available.

In a system such as Just a Line, many of these questions can be resolved by consulting the clients, Harry and Sue, about their preferences. This helps to ensure that the system looks as they want it to and also gives them the feeling of being part of the system development process. User involvement is beneficial in all areas of system development, and in designing the user interface it is essential. An interface that is quick to learn, easy to use and matches the user's view of the tasks in hand means not only a happier user but also fewer errors and a more efficient and effective user performance.

8.7 The Internet

If business goes well Harry and Sue may soon find that the most effective way of selling the Just a Line cards is via online shopping over the Internet. Even if they do not want to deal with online orders at present, they should at the very least have a website to market the cards and publicize the company. An example of the home page for a website for Just a Line is shown in Figure 8.32.

When designing for the Internet, developers should follow the principles of good screen design and bear in mind the issues mentioned in the previous section. In addition to this, there

Just a Line

offers you individual cards for every occasion.
Why not browse our extensive collection
of beautiful designs?

Birthdays - Anniversaries - Christmas - Easter - New Baby - Miscellaneous

We have many original designs that you won't find anywhere else.

From our site, you can choose the perfect card for every occasion.
You can even add your own personal message at no extra charge.

Winter market - one of our current popular designs

Follow the links to: View the cards Place an order Contact us

Figure 8.32 Example website for Just a Line

are other points to consider that relate specifically to website design. Some of these are listed below.

■ The home page is crucial, since this is where visitors decide whether to get involved in the site or not. A good home page allows visitors to see all the main areas of information and how to get to them.

■ A web page is not like a printed document; the most important information should be clearly visible at the top of the page, so that visitors do not have to scroll vertically or horizontally to see it.

■ The home page acts as a menu for the rest of the website, and should contain links to all the main areas that are easily visible. In Figure 8.32, for example, the Just a Line home page states clearly what the company's business is, gives an example of one of the cards, and provides links that allow visitors to view all the cards, place an order and contact the company.

- As shown in Figure 8.32, there should be only a few words on each link, so that it is clear to the visitor where to click.
- Links allow visitors to get around the site, but it is still important that there is an underlying structure that is easy to follow. Most websites use a hierarchical structure, which makes clear to visitors where they are in the site, what they can do there, and how they move forwards, backwards or return to the home page. One useful rule for website design is that it should never take visitors more than three clicks to get to the information they require.
- Most people are uncomfortable reading large amounts online, so text on the website should be kept as brief and to the point as possible. It should also be possible to print out documents, such as an order form for Just a Line cards, for customers who like to use the site to view cards, but want to order offline.
- *Multimedia* (the combination of different forms of media in a computer-based system) is one of the most exciting forms of user interface. A multimedia system may use text, graphics, sound, photographs and video, both separately and in varying combinations, so that information can be presented in a whole variety of ways. It is often tempting for website designers to include multimedia effects on the pages of the site in order to attract and entertain visitors. However, elaborate design can sometimes get in the way of the functionality of a website; visitors may simply be bemused or irritated by excessive graphics or animations. Multimedia is not always appropriate, particularly where visitors may have text-only browsers or are accessing the site via a modem. The home page in Figure 8.32 uses a thumbnail image to give visitors an initial view of one of the Just a Line cards. This can be linked to a large high-quality image of the card, which visitors can choose to view if interested, by clicking on the thumbnail.
- For a company such as Just a Line, the website is an important part of its marketing. This means that it is essential that the site is properly maintained and the information on it is kept fully up to date. It is also important for visitors to find the site easily, so it is well worth putting a lot of effort into choosing the right keywords that people browsing the web for products like the Just a Line cards are likely to put into a search engine.

Summary

In this chapter we have discussed different ways of implementing a system such as Just a Line, and looked at some of the issues a developer must bear in mind when implementing a small business system in a relational database package such as Microsoft Access. We gave a brief introduction to Structured Query Language (SQL), which allows users to manage data, get the answers to questions from the database and output the results. The chapter also looked briefly at how to design the user interface to a small business system, and how to create an effective and efficient website.

Exercises and topics for discussion
What can you remember?

You will find all the answers in the chapter

a) List two advantages and two disadvantages of programming a system from scratch for Just a Line.

b) List two advantages and two disadvantages of using a commercial package to implement the Just a Line system.

c) What is data redundancy and why is it a problem?

d) What is the role of a DBMS?

e) List three tasks that a user can carry out using a database package.

f) What is a database query?

g) What are forms and reports used for in a database package?

h) Which analysis model is the main source of information for the developer when constructing tables in the database?

i) Which analysis models are used by the developer when implementing the functionality of the system?

j) What is the user interface? Why is it so important for a system to have a user-friendly interface?

k) Give two reasons why the home page is a particularly important part of a website.

l) What is multimedia?

m) Why is it important to find the right keywords to put on a website?

8.1 Figure 8.33 shows the entity **Supplier** as identified in the entity-relationship model in Chapter 6. Design a database table to implement this entity and fill the first two rows with imaginary data.

8.2 Figure 8.34 illustrates a one-to-many relationship between the entity **Customer** and the entity **Order**. Explain how this relationship is implemented in a database.

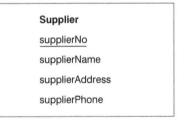

Figure 8.33 The entity Supplier

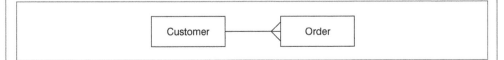

Figure 8.34 A one-to-many relationship between the entity Customer and the entity Order

8.3 In order to ensure that their deliveries are as efficient as possible, Just a Line has to organize the delivery routes very carefully. A delivery may include many customers, and a customer may be on many deliveries. This many-to-many relationship is shown in Figure 8.35; explain how it may be implemented in a database.

8.4 Think about the interfaces of systems with which you are familiar in everyday life – mobile telephones are one example. In a group, discuss whether the interfaces to these systems are satisfactory and how they might be improved.

8.5 Design a screen that will be used to enter details of an order for a Just a Line customer and then evaluate it in terms of the guidelines for good screen design listed in this chapter.

8.6 Figure 8.36 shows part of the E-R model for the Just a Line system (see Figure 6.26) implemented as linked tables in Access. If you are able to use an Access database system, set up tables and relationships as shown in the diagram.

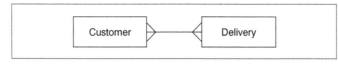

Figure 8.35 The many : many relationship between customers and deliveries

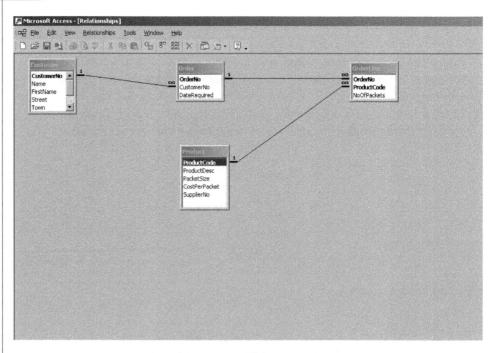

Figure 8.36 Microsoft Access database implementation

SQL exercises

The following 20 exercises relate to a small database system managing loans of DVDs. Figure 8.37 shows three of the tables from the database: DVD, DVDCustomer and DVDLoan.

DVD			
DVDNo	**DVDName**	**Category**	**DailyLoanRate**
D121	Closer	Thriller	2.50
D173	The Pledge	Thriller	2.00
D226	Stage Beauty	Historical	2.00
D297	Phantom of the Opera	Musical	3.00
D339	Notting Hill	Romance	2.50
D406	The Sun King	Historical	2.50
D554	The Madness of King George	Historical	1.50
D607	Psycho	Thriller	1.50
D661	Life of Brian	Comedy	1.50
D717	The Others	Thriller	3.00
D729	Sixth Sense	Thriller	3.00
D734	Finding Nemo	Animation	3.00

DVDCustomer					
CustNo	**CustFirstName**	**CustName**	**CustAddress**	**CustPhone**	**Status**
C046	Paul	Harman	27 Maple Road, Greenward	01822 770095	Adult
C102	Stan	Jakes	14 High Street, Greenward	01822 765803	Senior Citizen
C191	Pat	Smithson	23 River Road, Greenward	01822 609845	Adult
C288	Jane	Knowles	9 Green Lane, Greenward	01822 787855	Adult
C400	Jenny	Lewis	16 Oak Drive, Hansford	01823 577485	Senior Citizen
C487	Charlie	Jones	76 High Street, Greenward	01822 765092	Junior
C517	Jane	Davis	11 Lime Street, Hansford	01823 568449	Adult
C522	Jenny	Watson	32 Green Lane, Greenward	01822 763221	Junior
C534	Brian	Wells	46 Oak Drive, Hansford	01823 545329	Adult
C571	John	Burton	2 Elm Road, Greenward	01822 604233	Senior Citizen

(continued)

Figure 8.37 Tables DVD, DVDCustomer and DVDLoan

DVDLoan			
CustNo	DVDNo	DateIssued	DaysPaid
C102	D173	04/05/2005	2
C102	D554	04/05/2005	2
C191	D406	04/05/2005	2
C487	D734	01/02/2005	1
C517	D607	02/05/2005	1
C517	D661	02/05/2005	3
C517	D717	02/05/2005	3
C522	D339	02/05/2005	1

Figure 8.37 (Continued)

Write commands in SQL to do the following.

8.7 Insert details of a new loan for customer number C191, who hires DVD number D406 on 5 May 2005 and pays for two days.

8.8 Update the DVD table by increasing the daily hire rate of all DVDs by 2.5 per cent.

8.9 Delete details of the customer whose name is Brian Wells.

8.10 Display full details of all the DVDs.

8.11 Display the numbers and names of all DVDs in the historical category.

8.12 Display the different categories of DVD recorded in the system (each category should appear only once in the table).

8.13 Display the number of customers in the system in a column called Total.

8.14 Display the number of junior customers in a column called Juniors.

8.15 As a promotion, customers are given a free day each time they rent a DVD. Display the **DVDLoan** table with an extra column, called **ExtraDay** that shows the new number of days allowed for each loan.

8.16 Display the minimum (MIN) daily loan rate for a DVD. Display the answer in a column called **MinRate**.

8.17 Display the average (AVG) daily loan rate for a DVD. Display the answer in a column called **Average**.

8.18 Sort the DVD records in ascending order according to the daily loan rate.

8.19 Display the DVD customers' first names, names and phone numbers in alphabetical order.

In the following exercises you will need to use more than one table.

8.20 Display the names of the DVDs that are currently on loan.

8.21 Display the customer numbers and names of customers who currently have DVDs on loan.

8.22 Modify the query that you wrote as an answer to the previous exercise so that each customer's name only appears once.

8.23 Display the customer numbers and names of any junior customers who have borrowed a DVD in the **Animation** category.

8.24 Display the names of any DVDs that have been borrowed by senior citizens.

8.25 What is the query below asking? Write out the table that is the result of the query.

```
SELECT DVDCustomer.CustNo, CustName
FROM DVDLoan, DVDCustomer
WHERE DVDCustomer.CustNo = DVDLoan.CustNo
AND DaysPaid < 2
ORDER BY CustName ASC;
```

8.26 What is the query below asking? Write out the table that is the result of the query.

```
SELECT DVDname
FROM DVDLoan, DVDCustomer, DVD
WHERE DVDCustomer.CustNo = DVDLoan.CustNo
AND DVDLoan.DVDNo = DVD.DVDNo
AND Category = 'Thriller'
AND Status <> 'Junior';
```

References and further reading

Begg, C. and Connolly, T. (2004) *Database Systems: A Practical Approach to Design, Implementation and Management*, 4th edn, Addison-Wesley, Harlow.

Date, C.J. (2003) *An Introduction to Database Systems*, 8th edn, Addison-Wesley, Reading, MA.

Howe, D.R. (2001) *Data Analysis for Database Design*, 3rd edn, Butterworth-Heinemann, Oxford.

Shneiderman, B. and Plaisant, C. (2004) *Designing the User Interface: Strategies for Effective Human–Computer Interaction*, 4th edn, Addison-Wesley, Reading, MA.

Sommerville, I. (2004) *Software Engineering*, 7th edn, Addison-Wesley, Wokingham.

CHAPTER 9
TESTING AND HANDING OVER THE SYSTEM

In this chapter we look at:

- stages of system testing

- changeover methods

- system documentation

- maintenance

After studying this chapter and working through the exercises you will be able to:

- describe how testing is carried out at different stages of system development

- evaluate the appropriateness of a changeover plan

- explain what system documentation is required

- explain what is involved in the maintenance of a system

Introduction

This chapter concentrates on the final stages of a system development project, from testing through to installation of the system in the client organization. This includes a brief look at how testing is carried out and a discussion of different changeover procedures. We describe the documentation that must be supplied with the delivered system for the convenience of both users and technical staff. We also briefly consider what happens to a system after it has been handed over to the client.

9.1 Testing

Testing is often seen as a means of establishing that a software system is error-free and that it does what is required. It is virtually impossible to test a system so thoroughly that it can be claimed to be free of errors. Often, an attempt to correct one error may give rise to a host of others, which in turn have to be corrected and exhaustively tested. It is much more realistic to think of testing as a process of finding errors. When a stage is reached where the software appears to run perfectly, this does not mean that there are no more errors; it simply means that those errors have not been discovered.

It is hard even deciding how to establish that a system does what it is meant to. To test an information system like the Just a Line system, testers will typically be testing a set of programs against a requirements specification expressed in terms of informal diagrams such as data flow and entity-relationship diagrams. There is no way of proving conclusively that the program matches the specification, let alone that the specification reflects what the client originally wanted. Inadequate though it is, the testing process is still the principal tool available to developers to help them establish whether or not the system that has been built is what was required.

In the process of software development, testing has historically been left until the code has been written. However, testing or **quality assurance** activities can, and should, take place throughout the entire development process. In the early stages of development (i.e. at the analysis and design stages) the testing activities are designed to establish, as far as is possible, that the users' requirements have been thoroughly understood. The type of testing that can be done at this stage is of a different nature from the testing that is done during and after implementation. Before implementation, ideas are being tested; during and after implementation, code is being tested.

Pre-implementation testing

Pre-implementation testing is not done by programmers or even, normally, by members of a testing team, but by a team of reviewers who may be project managers, clients or system developers – anyone actively involved in the development process. Early testing is more often thought of as a review of work done so far – a walk-through of ideas or concepts. In the traditional life cycle approach this work will be done at the end of each of the development stages and will take the form of a review of the deliverables for the stage just completed. At the end of each stage the system development team must check back to the requirements specified in the previous stage (see Figure 9.1) and redo the work if necessary.

Reviewers check that the requirements specification document produced at the end of the requirements engineering stage accurately captures the client requirements and that the requirements specified are consistent and feasible. The set of alternative solutions produced at the design stage will be reviewed to check that they can meet the specified requirements, that they are feasible technically, operationally and financially, and that they would fit in with client or company policies and resources. In the technical specification produced at the end of the design stage it is still ideas that are being checked and not code, although these ideas are now more formally expressed in the design documents. One of the proposed solutions will have been selected and designed in detail. Reviewers will check that the design meets the requirements outlined in previous stages, that the design is complete and that it is compatible with logical and physical performance requirements.

The next set of tests will be carried out on the code and, with these, testing moves into the **post-implementation testing** phase. Initially this will be done by the programmer or a dedicated testing team.

Post-implementation or code testing

Most software developers perform testing of a system from the bottom up. First, small program units are tested individually, to check that they have been coded correctly. This is usually carried out by the programmer who coded the program. These units are then combined into program modules, which deal with specific parts of the system (examples from Just a Line are stock control, ordering and marketing). The modules are tested in the same way as the small program units, again usually by the programmer. When all known bugs have been eradicated, the system as a whole is thoroughly tested by the developers, a separate team of testers and eventually the clients.

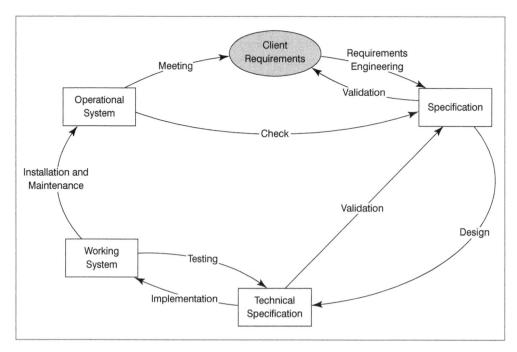

Figure 9.1 Simplified system life cycle

Whatever component of the software is being tested, the tester can apply two main testing techniques:

1. black box testing
2. white box testing.

Black box testing

With black box testing the tester is interested only in what the code does, not how it works. The code is tested by inputting data to the black box (program) and checking that the output is as expected. A program or unit is tested without knowledge of the internal structure of the code. Therefore this can (and should) be done by someone other than the programmer. The tester treats the code as a black box into which he or she cannot see. Test cases are selected on the basis of the externally visible behaviour of the code, as specified in the program documentation.

Black box tests of the code that produced the Just a Line invoice shown in Figure 8.5 would include a check that, if normal valid data were used:

- the invoice produced was for the right customer
- the customer number matched the customer details
- the correct prices had been picked up by the code
- the calculations were correct.

White box testing

White box testing, variously known as white box, structural, program-based or glass box testing, is done by someone who has access to the source code – usually the programmer. Test cases are selected to exercise those parts of the code likely to cause trouble – for example, branches, loops, boundaries, case statements and complicated algorithms.

White box tests of the code that produced the Just a Line invoice would additionally test the code with invalid output data. For example:

- invalid dates such as 30/02/05
- invalid customer numbers e.g. A, \, # (assuming that valid customer numbers were between 1 and 10 000)
- customer numbers that were out of range e.g. 200 000, –12
- customer numbers that were on the boundary e.g. 0, 1, 9999, 10 000, 10 001
- a blank customer number
- a blank number of cards.

White box tests would also check that the code could cope with extreme situations. For example:

- a customer order that included so many items that the invoice overflowed onto a second page
- a card name that was longer than the space allowed on the invoice
- an invoice that added up to an exceptionally large amount.

Test phases

Traditionally, code testing falls into three distinct phases:

1. unit testing
2. integration testing
3. system testing.

Unit testing

There is not total agreement over the definition of a unit. It has been variously defined as:

- a single cohesive function or procedure
- the smallest segment of code that can be separately compiled
- a function that fits onto a single page
- code that one person can write in a given amount of time.

The most commonly used definition is the first: that a unit comprises a single procedure or function. During unit testing a unit is tested by itself; any units that are called by it (server units) are replaced by stubs – short sections of code that substitute for missing units. A stub is required, for example, when the unit being tested calls another unit that has not yet been written. A dummy unit (the stub) is written, which, typically, does nothing except simulate a result or output a message such as 'Unit x has been reached'. This allows the unit undergoing testing to execute without the compiler or linker protesting about the missing units. The normal procedure is that a driver program will be written to act as a test harness for the unit being tested.

At unit testing level both black box and white box testing are appropriate.

Integration testing

Integration testing checks that units tested in isolation work properly when put together. When units have successfully completed the unit testing stage, they are combined into logically coherent groups for testing. For example, units forming a sub-system might be tested together. When testers are happy that a sub-system works, it will be combined with another sub-system for testing; when that combination works, another will be added, and so on, until the whole system is tested. The emphasis in integration testing is on checking that units work together; the testers concentrate on unit interaction rather than functionality. The code being tested gets progressively bigger – first a group of co-operating units will be tested together, then a sub-system, then several sub-systems; finally the whole system is tested.

System testing

During system testing, testers check that the system behaves correctly when used as the client might typically use it. At this stage, testing assesses whether the system can cope with mainstream data and conditions – the type of data users will expect the system to handle on a routine basis, under normal conditions. The tester will also check the system's performance behaviour – how long it takes to execute given tasks. The system will undergo volume, stress and storage tests to check its performance under extreme conditions – for example, huge volumes of input, high-speed input, large numbers of users, peak bursts of activity. All computations are checked for correctness with expected and unexpected data and conditions. The system's error handling and recovery will be tested to establish that the correct error messages are issued and that the system recovers gracefully from errors. The tester will also be expected to check that the system has an appropriate degree of security – that unauthorized users cannot gain access to it.

Other types of test

Other types of testing may be appropriate, depending on how essential it is to minimize software failure, and how much testing time and effort the client is prepared to pay for. The options include:

- regression testing
- acceptance testing by user or a testing team
- beta testing
- release testing.

Regression testing

It is a well-recognized phenomenon that correcting software faults very often leads to the introduction of new faults. The wise programmer therefore, when he or she has corrected a fault, should rerun all applicable tests and check that the program still produces the same results. This rerunning of tests and checking of results is known as regression testing. As regression testing is often a very tedious and expensive business, programmers are encouraged to organize their testing activities so that regression testing can be as automated as possible.

Acceptance testing

Acceptance tests are done by the client when the software is delivered. Normally, acceptance tests are a formal set of tests run to determine whether or not a system meets the client's acceptance criteria. This, in turn, will determine whether or not the client is prepared to accept and pay for the software. The main functionality of the system is tested, using the type of data a user might be expected to input. Acceptance tests are often circulated in advance so that the programmer or development team can run them before formally handing over the software. This can save everyone time and embarrassment.

Beta testing

If a software product such as a word processor, a web browser or a stock control system is being developed for general release, the software developers may decide to have it tested by outsiders before it is finally shipped. **Beta testing** is done by teams of 'friendly' outsiders who represent the type of users likely to purchase the software being developed. Their role is to give feedback on their experience of using the product in a working environment.

Release testing

Release tests check that the product about to be sent out is complete: that all the disks or CDs are there with the right files on them; that the correct versions of the files are being used; that the disks or CDs are virus-free and that the correct set of documentation is included. The tester will do a high-level check that the software does what it claims to do by comparing the software, the requirements documentation, the marketing material and the user documentation.

Types of test data

In each of the above groups of tests a program, or suite of programs, is executed with predefined sets of data to see if the program behaves as expected and produces the desired results. The types of test usually carried out fall into the following categories.

■ Valid data, which would be normal for the system. In the Just a Line system this would include entering details for a typical customer and an average customer order for some cards. Tests in this category also include boundary conditions, i.e. data that is not typical, but that is still within the correct range for the system. Examples of boundary conditions include the following:
 – a delivery charge of 0.0
 – an order for a very large number of cards
 – items of data that completely fill the defined fields, such as a long customer name or address.

■ Data that is invalid and should not be entered into the system. In Just a Line this could include:
 – a delivery date for 3/6/1892
 – an order for a card design that does not exist.

■ Data that is itself valid, but that is not acceptable to the system in its current state. Examples from Just a Line include:
 – trying to enter details for a customer who is already on the customer file
 – trying to amend an order that has already been supplied.

■ Data that tests the performance of the system, such as the speed of response and the handling of large volumes of data.

At the system testing and acceptance testing stages we need data that tests the overall system requirements. A useful basis for this testing is the requirements specification document or a separate problems and requirements list, which may have been compiled during the system development. This is a record of all problems and requirements that are mentioned by clients in interviews or that are subsequently discovered during analysis of the system. The list initially records each problem or requirement together with its source. During the development of the system, as each problem is dealt with, this is also recorded on the problems and requirements list. An extract from the problems and requirements list for the Just a Line system is shown in Figure 9.2. For each problem or requirement that has been identified, the developer should create a test that will examine whether or not the system solves the problem or fulfils the requirement.

Who does the testing?

The testing of the code may be done by different people at different times. Who does the testing will depend both on the stage at which testing is being done and the resources allocated to testing a particular software product. Testing may be done by:

- the programmer
- a team of testers
- beta testers – people who represent the market for the software
- the client
- the maintainer (often a programmer other than the one who wrote the code).

Unit testing is normally done by the programmer who wrote the unit. The programmer is not necessarily the best person to do this job as no one wants to find errors in their own code. Studies indicate a reluctance on the part of the developers to put much effort into trying to prove that their own work is flawed. Also, programmers are rarely trained testers – it is estimated that only 10 per cent of programmers are given adequate training. Nevertheless, resource limitations usually dictate that it is the programmer who does the unit testing.

Integration and system testing are often done by a testing team. The client will subject bespoke software to acceptance tests. Beta testing is done by potential users on non-bespoke software.

User trialling

In most development projects the system is handed over once the client is satisfied with the results of the acceptance tests, but sometimes a further period of more extensive testing may

Problem and Requirements List for Just a Line		
Problem or requirement	**Source**	**Solution**
1. No way of knowing who the regular customers are.	Sue	Set up a file of all customers who place at least one order each year.
2. Telephone orders are very time-consuming to deal with.	Harry and Sue	Provide written order forms that can be sent out with mail shots.
3. What is done with past orders?	Analysis – entity life history of order	Create a separate file for past orders and delete any orders more than one year old.
4. It would be nice to be able to see at a glance how various cards are selling.	Sue	Set up a spreadsheet recording sales for all cards and produce a bar chart that shows comparative popularity of the different cards.

Figure 9.2 Extract from the problems and requirements list for Just a Line

take place that is run entirely by the client. This is known as user *trialling* and differs from acceptance testing in the following ways.

- The whole process is run by the client and the other users of the system.
- Test data is supplied by the client and should be real data wherever possible.
- Large volumes of data are needed to create a realistic situation.
- Users of the system must have some training before trialling can be carried out.
- Trialling involves a lot of planning, both for the developer and for the client organization.
- Trialling involves a lot of time and effort on the part of the client and users of the system.

In spite of the extra effort involved, particularly for the client organization, trialling repeatedly proves to be well worth this trouble in that it frequently uncovers fundamental errors in the system that have somehow slipped through previous testing procedures.

Testing documentation

All testing should be scrupulously documented for the benefit of all involved. This serves both as a working document and as a record of achievement for the benefit of those developing the system, and as proof that the testing has been done for the client. As such it provides invaluable information for whoever has to maintain the system. The documentation should specify the test data used, the condition being tested, the expected result and the actual result.

9.2 Changeover to the new system

Once all the planned testing procedures have been completed satisfactorily, the system can be handed over to the client and the changeover from the old to the new system put into effect. There are several ways of carrying out the changeover; the method chosen will depend on such factors as the type of system that is being installed and the preferences of the client organization. The changeover will require careful planning on the part of both the developer and the client to ensure maximum efficiency and minimum disruption to the normal working of the client company. Many changeovers involve extra work at night or at the weekend, both for the developer and for those members of the client organization who are most involved in the new system.

Methods of changeover

Four common methods of changeover to a new system are listed below, together with an indication of the advantages and disadvantages of each.

1. Direct changeover (also known as cutover)

- The old system ceases to function and is completely replaced by the new system.
- This type of changeover is usually carried out at the weekend or during a holiday period.
- Direct changeover is often used in online and real-time systems, where it is not feasible to run the old and the new systems in parallel.

- Duplication of work for the client organization is minimized, since only one system is running at any time.
- The changeover must be very thoroughly planned, so that the new system can take over with a minimum of trouble and run smoothly from the start. This is very important, since there is no old system to fall back on if things go wrong. If the new system has teething troubles, this can prove to be a disaster for the client company.
- There are no results from the old system that can be compared to see how well the new system is operating once changeover has been completed.

2. Parallel running

- In parallel running, the new system is installed alongside the old and the two are operated together until complete confidence in the new system has been established.
- Parallel running is much less risky for the client than direct changeover, since the old system can simply continue running if problems arise.
- Some of the results from the old system can be compared with those from the new, so that the client can get an idea of how the new system is operating.
- Running two systems side by side creates a huge workload for the client organization. This means that in many companies parallel running would not be a practicable method of changeover.

3. Pilot running

- In pilot running the system is run for a time in one area only. In Just a Line, for example, orders from customers who live in the immediate locality might be run on the new, automated system, while all other customer orders would be handled manually as before. As Sue and Harry gain confidence with the new system, more and more customer orders will be handled by the computer.
- The advantages and disadvantages of pilot running are a combination of those for direct changeover and parallel running.

4. Phased implementation

- In phased implementation the system is installed gradually, so that the client and other users can become accustomed to one area of business at a time. In Just a Line the mail shot program could be installed at first, followed by the programs dealing with customer orders, stock control, supplier orders and finally the accounting procedures.
- In phased implementation part of the system can be (and often is) installed before the rest of the system has been completed. This gives the opportunity of valuable feedback from the client while the system is being developed. It also takes some of the pressure off the developer, since small parts of the system can be delivered when they are ready, without waiting for completion of the whole system.
- Phased implementation would probably be the most appropriate method of changeover for the Just a Line system, since it allows Sue and Harry to concentrate on part of the new system and to become competent in using the system in stages. As their confidence increases, the new automated system will gradually take over in all areas of their business.

9.3 Documentation

One of the most important parts of the software system handed over to the client is the written or online documentation that accompanies the programs. This documentation serves several purposes.

- It enables developers to understand how the system has been built, so that it can be maintained and modified as necessary.
- It gives instructions for the day-to-day operation of the system and peripherals such as printers.
- It introduces new users to the system and allows them to find their way around it.
- It provides information that enables experienced users to exploit the system to its full potential.

Documentation of the development of the system

The documentation for developers, who will maintain and modify the system, should provide a complete record of the system development process from the initial client request to the final implementation and testing. Many of the documents that should be incorporated will already have been produced as deliverables during the development of the system:

- the original problem definition
- models of the current and required systems that were created during requirements engineering, such as the data dictionary, data flow and entity-relationship diagrams, descriptions of the system design, including module charts and pseudocode
- details of hardware and software used
- fully commented code for all programs that form part of the system
- the complete set of test plans, test data and results for the system.

All these documents should contain extensive commentary, which explains the reasoning behind the decisions taken during the system development. The documentation as a whole should show clearly not only what was done but how it was done and why it was done in that way.

Most of the documentation described here will be produced in parallel with the development of the system, but it is important that it is not forgotten once the system has been handed over to the client. Any changes made to the system during its lifetime must be recorded so that an accurate picture is maintained of the current state of the system.

Documentation for the operation of the system

The style and content of the documentation that covers the day-to-day running of the system will vary according to the size of the system, its complexity and the expertise of those people who are to perform the operating tasks. In a large organization, where the system may involve many different networks, operating duties will probably be carried out by trained technicians. Their responsibilities may include routine maintenance of equipment as well as the daily tasks that ensure the smooth running of the system. The documentation in this case can assume a certain amount of technical knowledge.

With a small organization, such as Just a Line, it is the clients who are in charge of the day-to-day running of the system. They will need instructions for relatively simple tasks, such as putting more paper in the printer and making regular back-ups of files. Documentation in this case must be simple and clear; no technical knowledge can be assumed. Jobs that require specialist knowledge, such as fixing machines, should be covered by a maintenance agreement between the clients and the hardware suppliers. There should also be continuing contact with the developer of the system, who can advise on problems concerning operation of the system software.

Documentation for novice users

It is very important that clients and users who are not familiar with computers receive full training in the use of the new system before they have to operate it on their own. Once the system has been delivered and the client is left in charge, it is the documentation, along with the online help facilities, that provide the main support for users. Although the system developer will maintain regular contact with the client organization for some time after delivery, it is simply not practicable for users to consult the developer about every little query. For immediate solutions to the problems that arise, clients must be able to turn to the user documentation.

User documentation for any system must, above all, be easy to follow. The user manual is a reference document, not a novel that will be read through from beginning to end. Information that the user needs must be simple to find. This means that the manual must have a clear structure and a comprehensive index. For novice users the style of the manual is also important. Any technical terms used must be defined in layman's language and even very simple operations should be explained in full, preferably using diagrams. There should also be instructions on how to get out of trouble at all stages.

Novice users are not interested in the advanced features of the system. Their priority is to perform basic tasks, such as entering data and producing reports. The section of the user manual that is aimed at new users must concentrate on enabling them to get started, to gain confidence with simple operations and to get out of trouble quickly and easily.

Documentation for experienced users

Experienced users of the system will not need to refer often to the user manual or online help, and even then probably only to discover possible short-cuts for frequently used operations or to read about the more complex features of the system. The documentation at this level can assume that the user understands technical terms and is already familiar with the basic workings of the system.

9.4 The post-implementation review

If all goes well, the first formal meeting of the client organization and the whole development team after delivery of the system is the post-implementation review. This typically takes place about three months after installation, by which time the clients will have got to know the system and can report on how it is functioning. The post-implementation review is a useful occasion for the developer to discover what lessons can be learnt from this particular system project. During discussion with the clients the following questions will probably be raised.

- How satisfied are the clients with the hardware and software supplied?
- Were the clients happy with their degree of involvement in the development of the system?
- How are staff of the client organization reacting to the system?
- To what extent does the system fulfil the original client requirements?
- How is the system performing in a live situation? Are response times fast enough and does the system cope efficiently with large volumes of data?
- How has the system affected the client organization in terms of turnover and future staffing levels?
- Does the client have any ideas for future enhancements to the system?

9.5 Maintenance

It is tempting to view delivery of the completed system to the client as the end of the development process. The money is handed over, the client takes charge of the system and the developer moves on to new projects. However, this is very far from being the end of the story. In fact, the stage euphemistically known as maintenance frequently swallows up more time and money than all the other stages of the development process.

The blanket term maintenance can cover everything from fixing minor mistakes in coding to unravelling major misunderstandings about what the client wanted from the system. On top of this, maintenance is often used to describe modifications and extensions, which may be requested only after the client has been using the system for some time.

There are any number of reasons for a developer being called back by the client to change some part of the system. Some of these are listed below.

- The computer supplier may have brought out a new version of its hardware and is gradually decreasing its support for older models.
- There may be new versions of the software that was used to build the system and support for previous versions is no longer available. Systems built using packages may be particularly vulnerable to this problem.
- Changes may occur that are external to the system and over which the client and developer have no control, yet which necessitate a change in the way the system operates.
  For example, a change in the ordering procedures of the Just a Line supplier may necessitate changes in the Just a Line system.
- There may be errors that have become apparent only after the system has been in continuous use for some time.
- The interface of the system may not be acceptable to those people using it. The staff of the client organization may not feel happy with the new system and may be reluctant to give up their former methods of work.
- In getting to know the system the clients may have identified extra features that would be useful to the organization and that they would now like the developer to provide.

Whether it is a matter of maintenance or modification, and whether it is a major or minor change that is required, any work on the system after it has gone live is a complicated process. It is unlikely that the developers who worked on the original system are still available, since they will all have moved on to new development projects. This means that the people fixing the system will have to spend some time getting to know how it operates and how it was built. The quality of the system documentation (see Section 9.3) is of vital importance here.

If major changes to the system are needed, the whole development process may have to be repeated, but now the working system has to be preserved at the same time. It is not acceptable for an organization, which has become dependent on a computer system, to close down while alterations to the system are carried out, so the developers have to work round the operational system, causing as little disruption to it as possible. Changes should be implemented by means of a fully documented change control procedure. This includes forms detailing the required change (completed by the user) and the developer's estimate of the implications in cost and time.

All changes to the system, however small, have to be fully tested, installed and documented. Thus, even small changes involve a lot of work for the system developer. Moreover, it is often very difficult to predict accurately how much time, effort and money a particular change will entail, since an apparently minor change can cause a ripple effect in the system. Given all these problems, it is hardly surprising that the maintenance stage is the one that is least popular with system developers.

Summary

Although it is at the coding stage that a computer system seems to come to life, there is still much work to be done before the development process is completed. This chapter describes briefly the final stages of a system development project, from testing through to installation of the system in the client organization. There are several different ways in which the changeover to a new software system can be achieved and it is important that the most suitable method is chosen in each case. Other factors that influence how the system will be received by its users should also be carefully considered. One of the most important factors is the quality of the supporting documentation. Finally, the chapter looks at what happens to a system after delivery, the purpose of the post-implementation review and the particular problems of maintenance and modification.

Exercises and topics for discussion
What can you remember?

You will find all the answers in the chapter
a) What is the main difference between pre- and post-implementation testing?
b) What does white box testing test? What type of test data would be appropriate?
c) What does black box testing test? What type of test data would be appropriate?
d) What do we mean by a program *unit*?
e) What are the main differences between *unit* and *integration testing*?
f) What is being tested during *system testing*? What type of test data would be appropriate?

g) Explain the terms *regression testing*, *acceptance testing*, *beta testing* and *release testing*. Whom would you expect to be doing each of these tests?

h) Programs are tested using various types of test data. Explain the types used and give examples of each.

i) How does *user trialling* differ from *acceptance testing*?

j) Why is it important to document the testing of a system?

k) What should be specified in a test plan?

l) List different changeover methods and explain the main differences between them.

m) What documents would you expect to be included in the documentation of a system?

n) What are the main reasons for providing system documentation?

o) As a system developer, what types of question would you expect to have to answer at a post-implementation review?

p) What type of work should a system developer expect to do during system maintenance?

9.1 Write a test plan for a program written in a computer language with which you are familiar. The program is to receive as input a series of positive whole numbers, terminated by a full stop and output the largest number in the series.

9.2 Discuss which method of changeover would be suitable for the following systems:
- a system that runs a school library
- a system for a group of GPs
- a system for the delivery operations of a small dairy.

References and further reading

Bennett, S., McRobb, S. and Farmer, R. (2004) *Object-Oriented Systems Analysis and Design using UML*, 3rd edn, McGraw-Hill, London.

Pressman, R.S. (2004) *Software Engineering: A Practitioner's Approach*, 6th edn, McGraw-Hill Higher Education, London.

Sommerville, I. (2004) *Software Engineering*, 7th edn, Addison-Wesley, Wokingham.

CHAPTER 10
MANAGEMENT AND PROFESSIONAL ISSUES

In this chapter we look at:

- project management

- the jobs of the project manager

- planning the project

- bar and network charts

- monitoring and controlling the project

- software metrics and quality

- the law and professional issues

After studying this chapter and working through the exercises you will be able to:

- explain why project management is important

- write a task definition

- draw simple bar and network charts

- discuss the importance of high-quality software

- summarize the basic legal and professional issues relating to software developers

Introduction

Up to this point we have considered various activities that are part of the system development process. We have looked at frameworks for system development, at a variety of modelling techniques, and at how a system such as Just a Line may be implemented in a relational database package. However, we have not yet considered how the whole process is to be organized, monitored, controlled and evaluated.

Management of software system development has much in common with the management of projects in other areas, but at the same time the nature of software systems also demands specific qualities and expertise from the project manager. Software by its nature is intangible, you cannot see it, hear it or touch it. This makes it difficult to model and difficult to identify the source of errors when things go wrong. Moreover, there is no universally accepted, standardized software engineering process, and software engineering is still not recognized as a formal discipline in the same way as mechanical or civil engineering. For all these reasons, managing a software system development project of any significant size is a demanding and challenging role.

In this chapter we look at how a small project, such as the Just a Line system, might be managed. We introduce some of the techniques available to help the project manager and we consider briefly the concept of quality in the development of software systems. We also look at some of the legal and professional issues that concern the developer of software systems.

10.1 What is project management?

No project can run efficiently without some sort of management. Even something as apparently straightforward as preparing a meal of spaghetti requires organization, checking and control. The chef has to make sure that all the necessary ingredients are to hand, such as pasta, herbs and tomatoes. He or she must know the order in which the preparation and cooking should be carried out and must check to make sure that the whole process is proceeding as planned. If the pasta is cooked half an hour before the sauce is even started, the result will be an indigestible heap of soggy spaghetti.

If more than one person is involved, the problems of management become even more complex. On top of the tasks already mentioned, the head chef in our example has to assign various tasks to the team of under-chefs so that they do not end up with mountains of spaghetti and no sauce. The chef's role includes constantly encouraging and cajoling the team so that they produce their best possible work, acting as mediator if there are any clashes of temperament among team members and liaising with the waiters so that the right dishes of food are ready at precisely the right moment. It is no wonder that head chefs in large restaurants hardly get time to do any cooking.

Developing software systems is in many ways similar to any other kind of project. The basic progression, from a vague notion of the problem to a solution that can be interpreted by a computer, is fraught with potential difficulties. Without a firm framework of planning, monitoring and control, and careful handling of the development team, such a project can very soon disintegrate into complete chaos. Figure 10.1 shows how the development of the system and the management of the development project can be viewed as separate, but interrelated, entities.

Why do we need project management?

As can be seen from the above example of the chef, even an apparently simple sequence of actions needs a certain amount of management. In any project, somebody has to be able to plan what is to be done and to keep track of how things are going in relation to the plan. In all but the smallest projects, this process of monitoring is important in controlling the overall costs and benefits of the project. If work is running late, or if costs are beginning to escalate, it is essential to discover this as soon as possible and to be able to see how the project as a whole is going to be affected. The sooner a problem is uncovered, the sooner corrective action can be taken and the possible effects of the problem curtailed.

Separating the management side of a project from the technical development side means that one person, the project manager, is responsible for organizing, planning, monitoring and controlling the project as a whole. These tasks are recognized as time-consuming and laborious, but essential for the smooth running of the project. Establishing a well-defined role for the project manager means that there is a clear point of contact for clients and an obvious person to whom team members can take their problems.

Possible drawbacks

Project management is all about making the best use of resources, yet management itself takes up time, money and other resources. A skilled developer, who is acting as manager for a project, may have to spend a large amount of time on planning and administration instead of contributing to the actual development of the software system. This could be seen as a waste of technical expertise. The tasks of project management, such as producing a plan, monitoring progress and writing reports, are all expensive, in terms of both time and money.

Project management is not a magic wand. Many software systems are still delivered late, well over budget and often fail to fulfil the client's requirements. Nor is this a peculiarity of software systems; many other industries have the same problems. It is not surprising that there is a tendency to ask whether our present management methods are always worth the cost and effort put into them.

Who is involved in project management?

At the very simplest level, a system development project may involve only one person. Yet even here, that person is playing more than one role. If the system is for personal use, the same person represents the client, who has requested the system and who must be satisfied

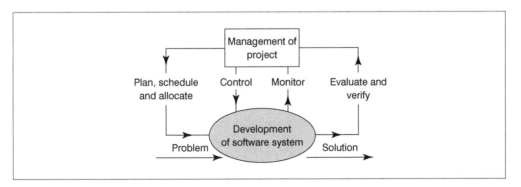

Figure 10.1 The development of the system and the management of the project

by the final product, the developer, whose task is to design and build the system, and the project manager, who plans, monitors and controls the progress of the system development.

In most situations these three roles are played by different people. Additionally, there may be many clients and users and a large number of software engineers in the development team. An important part of the project manager's job is to liaise between these various parties and to ensure that development of the system progresses as smoothly as possible.

The development team

Right from the start of a project, it is essential for a project manager to have the best possible team for the job. The technical expertise required will vary according to the type of system to be developed, but there are other, more general, qualities that a project manager should look for when selecting a team. Some of these are listed below:

- readiness to work hard
- ability to listen
- receptiveness to the ideas of others
- ability to communicate
- ability to co-operate with other team members
- enthusiasm for the project
- ability to cope with pressure.

The team must act throughout the duration of the project as a cohesive unit. A team member who does not fit in, however brilliant they may be technically, can seriously damage the project's progress.

The project manager

As well as having all the qualities already mentioned for team members, a project manager must be good at dealing with people in general. This is important, both for the management of the development team and the relationship with clients. The project manager is the link between the client and the team. It is the manager's job to keep the client up to date on how the project is progressing and to keep the team informed about the client's reactions to the developing system. When things go wrong, it is the project manager who must identify where the problem lies, explain the situation to the client and somehow reach a solution that satisfies everyone involved.

It is important for the project manager to have at least some technical expertise. This is necessary, first, so that the project manager is aware of and can appreciate the problems of the people developing the system and, second, so that the nature of the project is sufficiently well understood to identify the tasks and make accurate estimates of timings and costs. In a small project, the project manager will probably also be a member of the development team.

As well as technical skills, however, the project manager must have the ability to carry out other kinds of task that are specific to the management of the development project. A project manager must be able to:

- plan
- estimate time and effort (and therefore costs) as accurately as possible

- identify and set sensible tasks for the team
- use prior experience effectively
- schedule work
- use resources efficiently
- monitor and control the progress of the project
- evaluate what is produced and ensure high-quality work
- identify quickly the causes of problems and be flexible enough to adjust previous plans if necessary.

If we add to this list of requirements that the project manager should be good at handling people and should have enough expertise to participate in the work of the project, we can see that a good project manager really has to be all things to all people at all times.

10.2 Planning

If you look again at the list of project management skills above, you will see that several of these are concerned with making a plan for the development of the software system. Thus, a large part of the project manager's work has to be done before the actual development can get started.

Planning is basically a juggling act. A plan must reconcile what the system has to achieve with the resources available. Just as the objectives of the project will be different in every case, so the balance of resources will vary from one project to the next. Sometimes the client is prepared to spend plenty of money, but wants the system delivered within a very short time span; sometimes a developer with a particular type of expertise is not available; sometimes the client and users are unable or unwilling to spend much time in consultation about the development of the system. The project manager must have all this sort of information to hand in order to produce a workable plan.

It is important for the project manager to be able to identify areas of potential risk, such as inexperienced personnel, new technologies or problems relating to requirements, and to compensate for these by making contingency plans.

Estimating

The ability to estimate accurately is crucial to the success of the project plan. It is essential for the project manager and the client to have as clear an idea as possible of the time a project will take and the costs it will incur. There are at present no formal methods or rules for assessing the time or cost of a project in advance. However, there are certain techniques that can help the project manager to produce more accurate estimates. When estimating, the project manager should do the following.

- Make sure that all relevant information is available.
- Divide the project into small, well-defined parts before assigning times and costs.
- Be realistic; an estimate is not a deadline, nor is it an optimistic guess.
- Practise estimating and keep a tally of the results, in order to identify personal strengths and weaknesses. Some people are naturally better at estimating than others, but everyone can improve with practice.

- Reduce bias, by checking estimates with at least one other person.
- Make good use of historical data; this includes estimated and actual times and costs from previous projects, both from the project manager's own experience and from other people's.
- Distinguish between resource time and elapsed time. A job may require only one day's work on paper, but the time interval before it is completed may actually be much longer because of factors outside the project manager's control. This is particularly important in the situation where team members are working on more than one project at a time.
- Make extra allowance for contingencies, on the grounds that something is bound to go wrong.

Dealing with risk

Throughout the lifetime of the project, the project manager must deal with the risks that could threaten its successful completion. This means that the source of each risk must be identified, analysis carried out to determine the probability of the risk and its effect on the project, plans put in place to remove or minimize the risk, and regular monitoring carried out to check the current status of the project with regard to all the potential risks.

A whole variety of factors can threaten a project at any stage in development. The source of risks may include:

- technology – for example, relating to the capability or reliability of the software and hardware
- the client organization – for example, relating to changes in the client's management structure
- the project itself – for example, if key members of the development team are unavailable when needed.

Once the source of a potential risk has been identified, the project manager has to assess the likelihood of its happening and the effect this would have on the project. For example, it may be unlikely that there will be significant changes in the structure of the client organization during the course of the project, but new managers representing the client's interests in the project would cause considerable upheaval. On the other hand, it is highly likely that at some stage some members of the development team will be off with colds or flu, but as long as the project plan has taken account of this, it should not cause major problems.

Dealing effectively with risk means that contingency plans should always be in place in case the worst happens. For example, the project manager should always be in a position to produce a detailed and comprehensive report on the progress of the project and its contribution to the organization's goals. This means that if there is a change in client personnel, no time is lost in acquainting them with the purpose and current status of the project. In cases of illness or other unavoidable absence of team members, all members of the development team should have some knowledge of the work of other team members, so that someone can take over at short notice.

Identifying tasks

In the same way as software system developers decompose a problem to help them to understand it better (see Chapters 4, 5, 6 and 7), a project manager has to break down the project development into tasks and sub-tasks before the work to be done can be allocated.

This is where it is essential for the project manager to have a knowledge of the nature of the work to be carried out. It is here also that a detailed development method, such as those mentioned in Chapter 2, can be of great help to the project manager. A method prescribes the main decomposition of a system and the milestones in the development process. It also specifies the documents that are needed for each stage of development and the deliverables from each stage.

Examples of some of the high-level tasks that might be identified in the Just a Line development project are:

- produce a set of levelled data flow diagrams for the required system
- create an entity-relationship model of the required system
- produce user interface designs
- devise the test plan
- write user instructions.

A task should be as self-contained as possible, with a well-defined boundary. The definition of the task should include a clear statement of the objectives, the end product of the task, any constraints involved in carrying out the task and the resources required. It should also give an estimate of the time required to complete the task and room for the developer to insert the actual time taken. These figures are useful for estimating similar tasks in future projects. A definition for a task in the development of the Just a Line system is shown in Figure 10.2.

Task Definition	
Project	Just a Line
Date	July 2005
Task Title	Design document to be used as price list and order form for mail shot
Task Reference	JL-T15
Sub-Tasks	Further interview with clients to finalize details
	Design first draft
	Agree design with clients
End Product	Agreed design for price list/order form
Constraints	Design should not exceed 2 sides of A4 paper
	Order form section should be separate so that it can be torn off for use by the customer
Resources	1 team member (output design specialist)
	Design software, colour printer
	Clients available for consultation
Estimated Effort	1 developer for 1 day
Actual Effort	

Figure 10.2 Task definition for Just a Line

Determining availability of personnel and resources

The task definitions are an important part of the overall project plan, but on their own they do not provide sufficient information for the project manager to allocate and schedule the work to be carried out. It is also important to know exactly when team members and essential resources may be free for work on the project.

To record the availability of team personnel, charts should be drawn up that give details of each team member's weekly commitments and how much time is free to spend on tasks for this particular project. Developers often work on more than one project at any one time. In this case, the project manager may have to negotiate with managers of other projects for the time particular developers may be available. This, of course, greatly increases the complexity of the overall project plan.

As well as weekly commitments and other project allocations, the availability chart for each team member should give details and dates of any planned absences during the project, such as holidays and training courses. A typical personnel availability form for a member of the Just a Line development team is shown in Figure 10.3.

Scheduling

Once the project manager knows exactly what has to be done and the amount of time and resources available, it is possible to allocate tasks and schedule the work to be carried out. There are many automated project management tools on the market and they are of particular help in this area. They will produce a plan of work that aims to deliver the required system within specified time and budget limits, and that makes optimum use of the resources available. Automated tools are also invaluable when, inevitably, original plans have to be

Personnel Availability	
Project	Just a Line
Date	July–Sepember 2005
Name	Kate Simpson
Position	Designer
Weekly activities	Average hours per week (max. 40)
Work on other projects	7
Administration	1
Total non-project activities	*8*
Project management	0
Availability for project tasks	32
Planned absences	Dates
Training	4–5 August
Holiday	2–16 July

Figure 10.3 A personnel availability form for Just a Line

revised. No project ever goes entirely to plan; new tasks emerge, estimates turn out to be wrong, team members leave. Redrawing plans by hand can be a major headache, but automated tools considerably lessen the workload for the project manager.

The first step in the scheduling process is to identify milestones that will show that a particular phase of the project has been finished. In the development of the Just a Line system some of the milestones identified might be:

- specification of required system complete and agreed with clients
- interface designs agreed with clients
- testing complete
- user documentation written.

Milestones are essential for the project manager, not only in monitoring the project but also in reporting back to the client on how development of the system is progressing.

In the same way as a software system developer uses graphical models to help to understand and organize the problem space, a project manager uses graphics in the form of charts as an aid to scheduling work and making optimum use of resources. Although there are many different types of chart, they fall mainly into one of two categories: bar and network charts.

Bar charts

Bar charts (often referred to as *Gantt charts*) show the tasks to be carried out, together with the estimated and actual time for each task. They can also be used to show staff allocations to particular tasks and the dates they are expected to be working on them. This helps the project manager with the problem of scheduling staff who may be working on several projects over a period of time.

Figure 10.4 shows a section of a simple bar chart for the Just a Line project. The white bars show the estimated dates that a team member is expected to be working on a particular task and the shaded bars show the actual dates. We can see from the chart that Kate was allocated to task JL-T2 for the weeks beginning 8 July and 15 July, but in fact she completed the task before two weeks were up and so she was able to begin her next task (JL-T6) early. We can see that, at the time the chart was drawn up, this task was unfinished, but had already taken Kate longer than estimated.

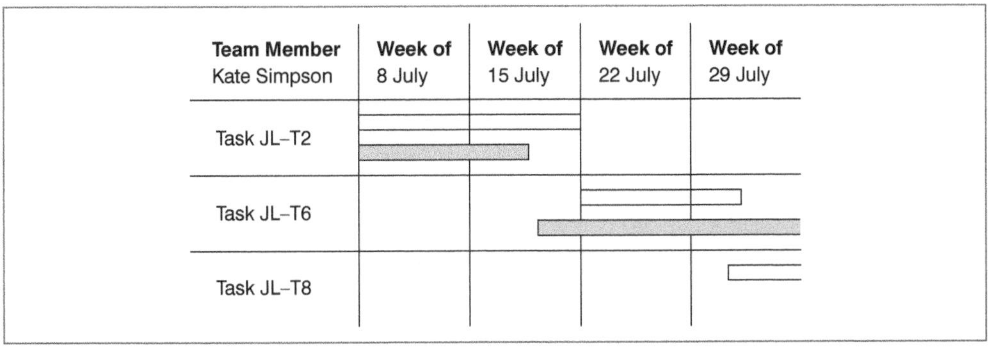

Team Member Kate Simpson	Week of 8 July	Week of 15 July	Week of 22 July	Week of 29 July
Task JL–T2				
Task JL–T6				
Task JL–T8				

Figure 10.4 A section of a simple bar chart from the Just a Line project

Task	Assigned to	Week of 8 July	Week of 15 July	Week of 22 July
JL–T1	MJD			
JL–T2	KS			
JL–T3	CHB			
JL–T4	CHB			
JL–T5	MJD			

Figure 10.5 A more sophisticated bar chart for Just a Line

Bar charts are a clear, convenient way of organizing and conveying information. They are the most popular at-a-glance way of monitoring projects, but they are limited in that they do not always give a good overall view of the project. On large projects it may be difficult to see from a bar chart how tasks relate to each other, or the effect on the project as a whole of a particular task taking much longer than estimated. Figure 10.5 shows a section of a more sophisticated bar chart, which includes the tasks carried out by all the team members.

Network charts

Network charts (or *activity diagrams*) use a notation that overcomes the limitations of bar charts. There are several variations on the basic idea of a network chart, of which the best known are PERT (program evaluation and review technique), CPM (critical path method) and CPA (critical path analysis).

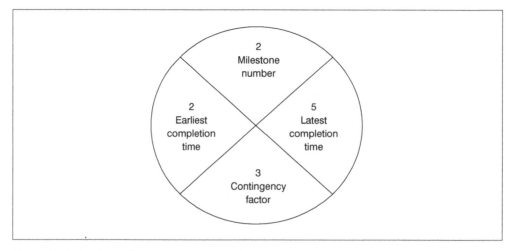

Figure 10.6 A project milestone as represented on a network chart

On a network chart, tasks are represented by lines with the estimated duration of the task written below each line. Circular symbols represent project milestones and are divided into four segments, as shown in Figure 10.6. The top segment shows the milestone number, the segment on the left indicates the earliest estimated completion time for the task (e.g. week number), the segment on the right gives the latest time the task can finish without affecting the overall project duration and the bottom segment shows the contingency factor (i.e. the difference between the two times).

A network chart shows how the various tasks are dependent on each other, which tasks must be performed in sequence and which in parallel. Figure 10.7 shows a simple chart for the design and implementation of a mail shot for Just a Line. Copies of the mail shot will be personalized and sent out to customers in the Just a Line database. The following tasks, with their estimated durations in hours, are represented in the diagram.

Task	Estimated duration (hours)
T1 – design personalized mail shot	4
T2 – agree mail shot design with clients	2
T3 – create small customer database for testing	2
T4 – write test plan	1
T5 – write code to produce mail shot	3
T6 – test	2
T7 – demonstrate to clients	1

In order to draw the network chart, we first need to work out the order of tasks: which ones can be started at any time and which are dependent on the completion of other tasks. The task dependencies are shown below.

Task	Must wait for completion
T1 – design personalized mail shot	–
T2 – agree mail shot design with clients	T1
T3 – create small customer database for testing	–
T4 – write test plan	T3
T5 – write code to produce mail shot	T2 and T4
T6 – test	T5
T7 – demonstrate to clients	T6

We can see, for example, that task T2 cannot be started until after T1, but that tasks T1 and T3 can be carried out before any other tasks have been completed.

Once the network has been drawn, the start and end dates for the project can be added to the chart. The project manager must be able to identify the *critical path*; this is the path for which all tasks must be performed on schedule so as not to cause delays in the project. Tasks that are not on the critical path can be delayed by the amount of time specified in the contingency segment without delaying the overall project. However, delays to tasks that are on the critical path will have an effect on the end date of the project, since the critical path is the path for which the contingency factor is zero for all tasks. The procedure for calculating the critical path is as follows.

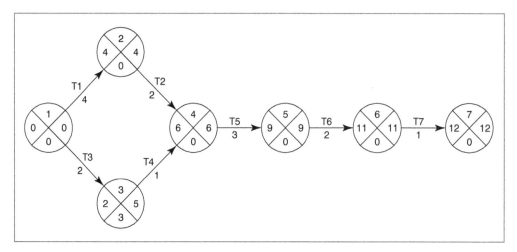

Figure 10.7 A simple network chart for the design and implementation of a mail shot for Just a Line

Forward pass: Starting with the first milestone, fill in the left-hand segment of each milestone, i.e. the earliest estimated completion time for the task leading to the milestone. For the first milestone, this will be zero. This is known as a forward pass and is carried out by adding the duration of the task to the number in the left-hand segment of the previous milestone. In the example in Figure 10.7, the number 4 (the duration of task 1) is added to 0 (the number in the left-hand segment of milestone 1) to give 4 as the number for the left-hand segment of milestone 2.

If there are two or more tasks leading to a milestone (for example, milestone 4 in Figure 10.7) calculate both estimates and use the larger number in the milestone. The figure in the left-hand segment of the last milestone indicates the earliest completion date for the project.

Backward pass: In the final milestone, indicating completion of the project, the left and right segments will always have the same figure, and the contingency factor will be zero. Starting from the last milestone and working backwards, fill in the right-hand side of each milestone, i.e. the latest completion time for each task. This is known as a backward pass and is carried out by subtracting the task durations from the numbers in the right-hand segments of the milestones. For example, in Figure 10.7 the right-hand segment of milestone 6 is 11; if we subtract the duration of task 6 (i.e. 2) from this we get the number 9 as the right-hand segment of milestone 5. If there are two or more tasks leading to a milestone (for example, milestone 1 in Figure 10.7) calculate both estimates and use the smaller number in the milestone.

Contingency factors: Calculate the contingency factor for each milestone; this is done by subtracting the figure in the left-hand segment from the figure in the right-hand segment. If the contingency factor on a milestone is greater than 0, this means that completion of tasks leading to that milestone can be delayed up to the contingency factor without delaying the project; in other words, there is some slack at that point in the project plan.

Critical path: All milestones that have a contingency factor of 0 are on the critical path; this means that the tasks on this path must be completed according to the estimated time or the project will be delayed.

Figure 10.8 shows the critical path for the design and implementation of the Just a Line mail shot. The critical path is shown by the heavy lines and shaded circles.

Both network and bar charts are used extensively in project management. One method of scheduling that aims to combine the best features of each type of chart is the network bar chart. This is simply a variant of a bar chart, but includes an indication of dependencies between tasks. In order to get a complete overall picture for the project plan, the project manager needs to make use of both network and bar charts, since each type of chart conveys different but essential information. In the same way as the various modelling techniques introduced in Chapters 4, 5, 6 and 7 each give the developer a different perspective of the system, the various types of scheduling chart give the project manager different perspectives on how the project is progressing.

The completed plan

The project plan, once completed, becomes the framework for the development of the software system. Individual project managers will have their own methods of documenting the plan, but this should incorporate all the following items:

- a summary of the objectives of the project
- network and bar charts showing scheduled start and completion dates, and a high-level view of allocation of staff and resources
- a detailed list of resource requirements
- a detailed list of estimated costs
- a statement of assumptions made in estimating, including risk assessment and contingency factor (extra time allowed for things to go wrong).

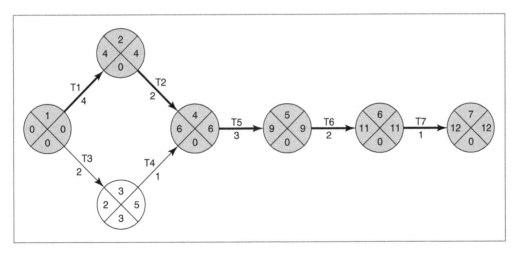

Figure 10.8 A simple network chart and the critical path

Appendices to the plan should include:

- detailed task definitions
- personnel availability forms
- detailed schedules.

Although it is important that the project plan is rigid enough to act as a framework for the development of the system, it is equally important that it is flexible enough to adapt to changing circumstances. Planning should be seen not as a once-and-for-all activity, but as an iterative process. Change is a constant factor in the life of a project; it comes from the client, from the development team and from external circumstances. Wherever it comes from, the project manager must be prepared to react to it in the way that brings the most long-term benefit, or the least long-term damage, to the project as a whole. This often means making amendments to the plan – extending the task list, rescheduling staff and resources, or recalculating costs. It is a critical part of the project manager's job to strike the optimum balance between firmness and flexibility in the management of the project.

10.3 Monitoring and controlling the project

Once work on the development of the software system is under way, the main task of the project manager is to monitor and control the progress of the project. To achieve this, it is important to be able to evaluate and measure how things are progressing in terms of time, cost and quality. At any given point the project manager should be fully aware of the following factors:

- the current status of the project with respect to the plan
- the cause of any deviations from the plan
- the corrective action that needs to be taken and the effect this will have on the project.

Good communication is essential for firm project control. There has to be some mechanism through which the project manager can keep a constant eye on how the system development is progressing. This mechanism often consists of a mixture of reporting by members of the development team, either at the end of each completed task or at regular time intervals, and a series of pre-planned meetings and *walk-throughs* at which the current situation is assessed and future action determined. During a walk-through, project team members may examine a specific deliverable, such as the user interface design for the system, as part of overall quality control.

What happens when things go wrong?

Corrective action, if needed to put the project back on course, must be taken as soon as possible. The sooner deviations from the plan are spotted and action taken, the more likely it is that long-term damage to the project will be avoided. Corrective action can take various forms, the most appropriate of these depending on the particular problem under consideration.

In any specific case the options open to the project manager include the following.

- The completion date can be put back so that the project goes late.
- Team members can work overtime. More resources, such as extra system developers, can be introduced to work on the project. This is an attractive, but potentially problematic, alternative. Not only can it dramatically increase project costs, but the need to inform new team members about the nature of the development, the current status of the project and the team's methods of working can use up the extra resources that have been added and seriously disrupt continuity and team spirit.
- The overall aims of the project may be changed. This could mean that some of the original requirements are omitted, or that the final implementation is less ambitious than originally intended. In an extreme case, the project manager may have to consider scrapping the project altogether.

When things go wrong with a software project there is rarely an easy or an obvious way out. The option chosen will always represent a compromise between reduced benefits from the system and extra costs incurred in terms of money, time and effort.

10.4 Software metrics

We have already said that it is essential to keep firm control of a project and to be able to evaluate how it is progressing, but we have so far said little about how this control and evaluation can be achieved. How can the project manager tell if the project is falling behind schedule? How is it possible to ascertain if the work produced is of the required standard? What exactly is a quality system?

A body of knowledge that aims to help software developers and project managers with these problems is *software metrics*. Software metrics are measurements that quantify aspects of software, both for the development process and the final product. They can be used both to evaluate what has already been accomplished and to predict what can be achieved in the future.

The most successful software metrics to date are those that relate to the management of the software development process. Examples of these metrics are measurements of elapsed time, the effort expended on a particular task and the cost of different activities. These measurements can be used as an aid to refining present plans and scheduling future projects.

Because we have mature and generally accepted metrics and accounting procedures relating to time and cost, it is relatively straightforward to assess the success or failure of a project from these points of view. Where quality of individual processes and products is concerned, however, this is a much more complex task. Although metrics for programs and specifications do exist, there are no universally agreed ways of assessing the quality of the software process at any given stage in development, or of predicting the quality of the final product. It is very hard to find meaningful metrics for something as intangible as software and to interpret them in such a way that they will accurately predict how good a software system will be. Measurements that are often recorded include the complexity of the original problem, the number of lines of code in the final system and the number of faults in the system reported by the client. In practice, however, such measurements alone are a poor indication of the nature of the final system. It is hard enough to define what we mean by a quality system, let alone attempt to measure it.

10.5 Standards and quality assurance

As software systems continue to play ever larger and more central roles in the world today, it is increasingly important for us to be able to produce systems of high quality. In order to do this we need first to be able to define accurately what we mean by a quality system. Unfortunately, this is much harder than it appears. To begin with, are we talking about quality from the point of view of the client or the system developer? For the client, quality means getting a system that does what is required of it, on time and at a reasonable price. The client will be interested in what the system can do, how user-friendly it is, whether it is reliable and how much it costs. The system developer will also be interested in these attributes, but on top of these he or she will be concerned with issues such as good program design, efficient use of resources and a well-structured database. Building a system is not like building a car or an aeroplane. Each system is individual and each will have its own set of quality criteria – in one case security of data may be of prime importance, whereas in another efficient use of resources may have top priority.

There are, of course, certain attributes that any good system should have. It should, for example, be reliable, easy to use, easy to maintain and, above all, it should satisfy clients and users. Standards do now exist, such as *ISO 9001*, which aim to capture these general attributes of software quality. It is now possible for organizations developing software to register conformance to ISO 9001 and indeed this is mandatory on many government projects. Another way in which companies can show that they are able to produce and deliver high-quality software is the *Capability Maturity Model (CMM)*. CMM originated in the USA, where the Department of Defense wanted a framework to help assess the capabilities of software companies that were bidding for defence projects. CMM focuses on how projects are managed and organized; it identifies five levels of maturity, where level one is chaotic and level five is a company that can demonstrate continuous process improvement. Companies that want to improve their processes can use CMM to assess their current level. In addition, potential buyers of software will find CMM levels helpful when comparing companies.

Quality systems do not come cheap and it is essential that everyone concerned with the development of a system participates in the drive for excellence. In many large software organizations there is an independent quality assurance division, which is responsible for assessing the software system. The quality assurance team on any project is sometimes viewed by the system developers with a degree of suspicion and even hostility. This is unfortunate, since they are there to help the developers, not to act as a police force. It is in the interests of everyone working on the project that a system of the highest quality possible is produced. There may also be a department that handles the *configuration management* of the system. Configuration management is responsible for documenting all modifications to the system and keeping track of all versions of it. This ensures that quality is maintained throughout any changes that the system may undergo in its lifetime.

In a small project, such as the Just a Line system, the project manager will have overall responsibility for quality assurance. This will involve careful monitoring and control of the development process and insisting on work of a high standard from all members of the development team. Each member of the team should feel responsible for the quality of his or her own work, and team members should be encouraged to review each other's work in a critical, but positive, way.

The amount of time and effort put into quality assurance depends on the nature of the system. *Safety-critical systems*, where the consequences of failure are life threatening, require the most rigorous quality assurance procedures. Examples of safety-critical systems that need this level of quality assurance are monitoring of hospital patients, flight control and nuclear plant control. This type of system may entail a justifiably high quality assurance cost, which may equal or exceed the development cost of the system. In contrast, for a standard information processing system, such as Just a Line, the cost of an appropriate level of quality assurance is likely to be a small proportion of the total development cost.

Structured walk-throughs are one of the most useful and effective ways of checking quality. These involve formal meetings as each milestone in the project is reached. During the walk-through the developer who is responsible for the material being reviewed presents the completed work to a panel of peer reviewers, including the walk-through organizer and a system developer who can judge the technical merit of the work. If appropriate, clients and users may also be present. Frequent walk-throughs during the project are an effective way of ensuring validation and verification of the developing system. *Validation* is concerned with matching the system to the customer's requirements. It asks questions such as 'Is the right product being developed?' and 'Will the system be fit for the purpose for which it is intended?' *Verification*, on the other hand, is concerned with issues of internal correctness – 'Does the system match its specification?' and 'Is the system being developed in the right way?'

Apart from walk-throughs, sound development methods, formal specifications, comprehensive documentation, rigorous testing and independent audits all help to ensure that the final system is indeed a quality product. It should also be remembered that building quality in to a system is only effective when carried out in conjunction with the people who are going to use the system. Users should be asked to specify what they consider as acceptable quality for the system interfaces. At the very least, users should be encouraged to scrutinize all input screen designs and output documents.

Software developers spend far too much time today on correcting mistakes that should never have been allowed to occur in the first place. Putting the emphasis on quality means trying to get it right first time round. Quality costs money, but the consequences of ignoring quality assurance issues can make all the difference between a system that is fit for purpose and one that falls short of requirements.

10.6 Legal and professional issues

A professional is an expert in a field that most clients know little about, and so they have to rely on the professional's experience and expertise. The work of the computer professional may affect the lives of large numbers of people in terms of their health, safety, finance or privacy simply because these people use a piece of software that the computer professional helped to develop. Just like professionals in any other discipline, practising system developers have to be aware both of the laws that are relevant to their work and the *codes of conduct* produced by professional bodies such as the British Computer Society (BCS) in the UK and the Association for Computing Machinery (ACM) in the USA.

Acts of Parliament

In the UK there are three Acts that directly impact on software professionals:

1. the Data Protection Act 1984 (amended in 1998)
2. the Computer Misuse Act 1990
3. the Regulation of Investigatory Powers Act 2000.

The **Data Protection Act** regulates how personal information (this includes both facts and opinions about an individual) is stored and processed on computers. It relates to any person or organization that holds and processes information concerning individuals. An individual about whom data is held has a right to receive a copy of the data, to know why it is being held or processed, and to be told details of any possible recipients of the data. An individual also has the right to prevent the data being processed if this is likely to cause harm.

The act is based on eight principles of good practice as shown below, all of which are enforceable.

1. Data must be fairly and lawfully processed; this means that the person who is the subject of the data must have given consent, or that the data processing is necessary for legal or other reasons, such as to protect the vital interests of the subject.
2. Data must be processed for limited, specific purposes; for example, medical records or university admissions.
3. Data must be adequate, relevant and not excessive; for example, a subject's salary is not needed in his or her medical record.
4. Data must be accurate and kept up to date.
5. Data must not be kept for longer than necessary; for example, data necessary for a research project must be destroyed when the project comes to an end.
6. Data must be processed in accordance with the rights of the person who is the subject of the data.
7. Data must be secure; the person or organization holding the data is responsible for protecting it against loss or damage.
8. Data must not be transferred to other countries without adequate protection.

There are further, more stringent conditions, relating to data that is defined as sensitive; this includes data relating to the racial or ethnic origin of the subject, their religious beliefs, their physical or mental health, their sexual life, and details of any offence committed, or alleged to have been committed, by them.

The Computer Misuse Act relates to offences concerning unauthorized access to computer programs and unauthorized modification of programs or data. The act covers accessing, or attempting to access, a computer without authorization, including via a network. It includes activities such as hacking, blackmail using emails, spreading viruses and modifying software (for example, a bank account) for personal gain.

The three main offences created by the act are as follows.

1. Unauthorized access to computer material; this carries a penalty of up to six months' imprisonment and/or a fine. Just looking at the data on someone's computer counts as an offence, even if no damage is done.

2. Unauthorized access to computer material with intent to commit or facilitate commission of further offences; this carries a penalty of up to five years' imprisonment and/or a fine. This is more serious than the first offence since it includes the intent to commit a further offence once access to the computer has been obtained.

3. Unauthorized modification of computer material; this carries a penalty of up to five years' imprisonment and/or a fine. This is the most serious of the three offences; it covers activities such as wilfully introducing viruses, altering data and deleting files.

The Computer Misuse Act also covers offences of conspiracy to commit or incitement to commit any of the three main offences.

The Regulation of Investigatory Powers Act regulates government's access to and surveillance of electronic communications. It gives the government considerable powers, including the right to obtain access to an individual's communications without the individual being aware of it, and the right to monitor how an individual uses the Internet. The act also allows employers to monitor employees' emails. This is a very complex piece of legislation and, in some cases, may conflict with the Human Rights Act 1998.

In addition to these three acts, there are many others that were not designed specifically for software systems, but that nonetheless apply to software professionals and their work. In the UK these include the Sale and Supply of Goods Act 1994, which states that goods must be of satisfactory quality and fit for the purpose for which they are sold, the Consumer Protection Act 1987, which gives a consumer the right to make a claim for injury or damage caused by a faulty product, and laws relating to copyright, which prevent developers from using other people's designs and code without permission.

Professional organizations

Software developers are supported in dealing with issues of law and professional practice by professional organizations such as the British Computer Society (BCS) and the Institution of Electrical Engineers (IEE) in the UK, and the Association for Computing Machinery (ACM) and the Institute of Electrical and Electronics Engineers (IEEE) in the USA. These organizations produce codes of conduct that set out standards and act as guidelines for software engineers in issues relating to ethical and professional behaviour. The BCS code covers four main areas.

1. Public Interest, which includes awareness of relevant legislation and the need to consider basic human rights.
2. Duty to Employers and Clients, which concerns quality of work and responsibilities to clients.
3. Duty to the Profession, which covers maintaining the reputation of the profession and supporting other professionals.
4. Professional Competence and Integrity, which includes conformance to quality standards and continually updating knowledge and skills.

The codes of conduct of the other professional organizations cover similar material, but are divided under different headings – for example, the ACM code has four sections: General Moral Imperatives, More Specific Professional Responsibilities, Organizational Leadership Imperatives, and Compliance with the Code.

The Ten Commandments of Computer Ethics

Computer Ethics Institute

Thou shalt not use a computer to harm other people.

Thou shalt not interfere with other people's computer work.

Thou shalt not snoop around in other people's files.

Thou shalt not use a computer to steal.

Thou shalt not use a computer to bear false witness.

Thou shalt not use or copy software for which you have not paid.

Thou shalt not use other people's computer resources without authorization.

Thou shalt not appropriate other people's intellectual output.

Thou shalt think about the social consequences of the program you write.

Thou shalt use a computer in ways that show consideration and respect.

Figure 10.9 'The Ten Commandments of Computer Ethics', created by the Computer Ethics Institute

You can find more information about all of these professional bodies on their websites, listed in the further reading section at the end of this chapter.

As the world of technology and computers becomes increasingly complex, there has been a corresponding rise in interest in how computer professionals should behave in different situations. There is a huge amount of material on the World Wide Web – for example, on the site of the Computer Ethics Institute, which has created 'The Ten Commandments of Computer Ethics'. These are reproduced here in Figure 10.9.

In the exercises that follow there are some short scenarios that you can use as a basis for discussion about legal, professional and ethical behaviour.

Summary

All system development projects need to be managed if the system is to be delivered on time, within budget, and if the best use is to be made of available resources. Together with technical excellence, good management is one of the keys to ensuring that the software produced at the end of the development process is of a high quality.

The job of the project manager involves a wide variety of tasks – from the technical, such as scheduling resources and monitoring progress, to the personal, such as dealing with the problems of individual team members. Today, automated tools and software metrics can alleviate many of the tasks of project management, but it is still the project manager who has overall responsibility for the efficiency of the development process and the quality of the finished system.

All practising software developers should be aware of the legal and professional issues relating to their work. One of the principal ways in which they can be supported in this is through professional bodies such as the BCS in the UK and the ACM in the USA. The codes of conduct published by these organizations set out standards and act as guidelines for software engineers in issues relating to ethical and professional behaviour.

Exercises and topics for discussion
What can you remember?

You will find all the answers in the chapter

a) Why is it important to monitor the progress of a software project?

b) What are the qualities that a project manager looks for when selecting his or her team?

c) Why does the project manager need technical expertise?

d) What does a task definition include?

e) Why are milestones important in project management?

f) What is shown on a bar chart?

g) What is the critical path on a network chart and why is it important?

h) What can the project manager do when things go wrong?

i) What are software metrics?

j) How can the ISO 9001 standard and the Capability Maturity Model help software companies?

k) What are the three principal acts that relate to computer use?

l) Name two other acts that a computer professional should be aware of.

m) What is a code of conduct?

n) What are the four main areas of the British Computer Society's code?

10.1 'I believe that large programming projects suffer management problems different in kind from small ones' (Brooks, 1995). Discuss the particular problems that might arise in the management of a large project.

10.2 Does the development of a small system, such as Just a Line, need to be managed?

10.3 Imagine that you have been seconded to work on the Just a Line project. Draw up your own personal availability chart for the next three months.

10.4 You will see below a list of task numbers, with their estimated durations and dependencies. Use this information to draw a network chart. Work out the critical path and list the tasks that lie on it.

Task	Estimated duration	Must wait for completion of
T1	6	–
T2	3	–
T3	1	T2
T4	2	T1 and T3
T5	1	T4

10.5 You will see below a list of tasks for making a meal of pasta with tomato sauce, together with their estimated durations and dependencies. Use this information to draw a network chart. Work out the critical path and list the milestones that lie on it.

Task	Estimated duration	Must wait for completion of
T1 – boil water	3	–
T2 – cook pasta	12	T1
T3 – fry onions	5	–
T4 – cook tomatoes	5	T3
T5 – add herbs and seasonings	1	T4
T6 – serve pasta and sauce	2	T2 and T5
T7 – grate parmesan	2	–
T8 – sprinkle parmesan over pasta	1	T6 and T7

10.6 Choose a small project that you will carry out on your own. For example, you could aim to write a program to add two numbers together in a programming language that is new to you.

(a) Make a list of the resources you will need for the project.

(b) Draw up a list of sensible tasks for this project and estimate how long each task will take you.

(c) Determine the order in which you will perform the tasks and see if it is possible for any to be carried out in parallel. Draw simple bar and network charts to help with this.

(d) Carry out the project, modifying your original plans where necessary.

(e) How accurate were your estimates? To what extent were you able to stick to your original plan?

10.7 Look at the code of conduct for one or more of the professional organizations mentioned in this chapter, and the material in Section 10.6. Use this as a basis for discussing how a computer professional should behave in the following situations.

The website addresses for the organizations are listed in the further reading section at the end of this chapter. You will find that this type of exercise is much more interesting and useful when carried out in a group, rather than individually. In each case you should identify the people and organizations involved and their rights in this situation. You should then list the different actions that are possible and, for each action, identify the benefits, risks and problems that arise from it. You should remember that there is generally no single right answer.

(a) A colleague at work is off sick and you are unable to contact him. Your manager tells you to copy some files from your colleague's computer so that you can work on a project that is running late. While doing this, you notice that your colleague has downloaded a large amount of material from an extreme right-wing organization. What should you do in this situation? Were you right to access your colleague's computer?

(b) You are a member of a team developing a mail order system. You are concerned that part of the system is not being properly tested, but your boss says that the tests carried out are adequate. Should you take further action or just forget about it?

(c) You work for a company that is developing a new system for a hospital. The system will monitor and dispense medication to patients in intensive care. You are concerned that there is a problem with the system that could put patients at risk and you think that some earlier design decisions need to be reconsidered, but your manager does not agree with you. What should you do?

(d) You are developing a new software system for a large client organization. As part of your job, the client asks you to check how the staff are using email and the Internet. Should you carry out the client's request or is it an invasion of staff privacy?

(e) You are part of a team developing a large software system, but you have got very behind. You remember that you still have some code written by a colleague who left some months ago, which you could use directly in the program you are writing. You do not know where your colleague is working now. Would you be wrong to use the code?

(f) You are a member of a team developing an accounting system for a large client organization. You have to give a presentation to the client on the progress of the project, but you are concerned because of some potentially serious problems that could delay completion. Your manager tells you not to mention this because the developers can probably sort things out and the people on the client team do not understand software projects, but you are not convinced that he is right. What should you do?

References and further reading

Association for Computing Machinery (ACM). See http://www.acm.org/.

Baase, S. (2003) *A Gift of Fire. Social, Legal and Ethical Issues for Computers and the Internet*, 2nd edn, Prentice-Hall, Upper Saddle River, NJ.

British Computer Society (BCS). See http://www.bcs.org/.

Brooks, F.P. (1995) *The Mythical Man Month*, 2nd edn, Addison-Wesley, Reading, MA.

Institute of Electrical and Electronics Engineers (IEEE). See http://www.ieee.org/.

Institution of Electrical Engineers (IEE). See http://www.iee.org/.

Kaposi, A.A. and Myers, M. (2001) *Systems for All*, Imperial College Press, London.

Pfleeger, S.L. (2001) *Software Engineering, Theory and Practice*, 2nd edn, Prentice-Hall, Upper Saddle River, NJ.

Pressman, R.S. (2000) *Software Engineering, A Practitioner's Approach*, 5th edn, European adaptation by Darrel Ince, McGraw-Hill, London.

Sommerville, I. (2004) *Software Engineering*, 7th edn, Addison-Wesley, Wokingham.

CHAPTER 11

CASE TOOLS AND ALTERNATIVE APPROACHES TO DEVELOPMENT

In this chapter we look at:

- CASE tools

- the role of prototyping in system development

- Rapid Application Development

- Agile Methods

- open source software

- object-oriented system development

After studying this chapter and working through the exercises you will be able to:

- describe the main features and role of CASE tools

- explain how prototyping is used in system development

- describe the main features of Rapid Application Development, Agile methods and open source development

- describe the main concepts of object-oriented system development

Introduction

The standard structured approach to developing software systems that we have described in this book has been around for many years and has stood the test of time. For many systems, this is still an efficient and effective development method. Today there are, however, very many different types of system, and it is no longer feasible or appropriate to use the structured approach in every case. Developers today have to be aware of other approaches in order to make the best choice of development method for any particular system. In this chapter we look at other approaches to system development: prototyping, Rapid Application Development (RAD), Agile Methods, open source and object-orientation. We begin by briefly introducing the topic of CASE tools, which are essential for consistency checking in larger systems, as well as providing support for the whole development process.

11.1 CASE tools

In a large development project, it is simply not possible to develop the system manually, drawing all the diagrams and checking consistency between them by hand. In this situation, developers have to rely on a **CASE** tool. CASE stands for Computer Aided Software Engineering and, like so many words in computer science, it incorporates a wide range of meanings. At the most general level a CASE tool can be any piece of software that assists in the development of a system. This can include compilers, debuggers and any automated design aid, such as a program that draws lines and boxes on the screen. At the top end of the range, however, the most sophisticated CASE tools include complete sets of integrated programs, which guide the developer through the whole process of building and managing a system. Some CASE tools are specifically tied to a particular development method, but many others claim to be more general and to fit in with a number of approaches to system development.

CASE tools and structured techniques

CASE tools automate the system development process, including modelling techniques such as those that we looked at in Chapters 4, 5, 6 and 7. Using a CASE tool a developer can, for example, quickly build up a data flow or entity-relationship diagram from boxes and other symbols provided on the screen. Hard copies of diagrams drawn with the aid of a CASE tool do not appear very different from those drawn carefully by hand or with a simple drawing package. Unlike a human, however, the CASE tool remembers all the details of every diagram that it has drawn.

Let us imagine that the developer has drawn the Just a Line context diagram (see Figure 4.4) with a CASE tool and now wishes to explode the diagram and move down to level 1. All the inputs and outputs from the context diagram are recalled and automatically produced on the screen so that they can be incorporated into the new level 1 diagram. This will be constructed by moving the symbols provided by the CASE tool into the required position on the screen, adding data stores and internal data flows between processes, and labelling all the symbols on the diagram. As the developer draws the level 1 diagram, the tool automatically checks that the work is consistent with any previous diagrams for the Just a Line system. When the developer wants to move on to build the level 2 data flow diagrams, the

CASE tool will explode each process, showing the relevant inputs and outputs. If the developer wants to make a change to one of the data flow diagrams later on, the CASE tool will register the effects of the change and automatically update the relevant diagrams. It does not allow the developer to infringe the balancing rules described in Chapter 4.

As well as checking consistency between different levels of diagram, the CASE tool can also create links and provide cross-checks between different types of model, such as data flow and entity-relationship diagrams. The CASE tool's repository (a huge, automated data dictionary) forms the basis for a whole network of connections between all the different models of the system. This sort of cross-checking, to ensure consistency between all the models of the system, is a major feature of all CASE tools, and can save the system developer many hours of tedious and tiring work.

Some of the more sophisticated CASE tools use models of the required logical system to generate code, database files and menus automatically. Entities in the data model become database files, attributes become fields, process names in top-level data flow diagrams become top-level menus and lower-level process names become lower-level menus. This means that a CASE tool can assist and support the developer throughout the entire development process.

CASE tools and the client

There is no doubt that CASE tools can be a great help to the system developer, but how does this affect the client and the quality of the final system? As we mentioned in the previous section, a hard copy of a diagram drawn with a CASE tool does not look very different from a similar diagram drawn by hand. It will, however, be very much easier for the developer to make any changes the client suggests to the CASE diagram, since it will not be necessary to redraw the whole diagram each time. The client may also be shown diagrams on the screen, in which case the developer can often make the required changes then and there. Diagrams such as these, which can be altered on the spot, are extremely useful for experimenting with ideas about the design of the system: 'What would happen if part-filled orders were stored in a separate file?', 'Suppose we combine the processes of ordering and control of stock?'

With a live diagram on the screen it is relatively easy to try out ideas and to see the effect various changes would have on the system as a whole. Used in this way, with a close partnership between the developer and the client, CASE tools are invaluable in *prototyping*, since they greatly facilitate fast production of prototypes. CASE is particularly useful in the rapid development of evolutionary prototypes where the working model is progressively refined to become the final system. A discussion of prototyping can be found in the next section of this chapter.

Understanding CASE

The whole area of CASE is shrouded in a blanket of terminology: you can hardly read even two paragraphs on the subject without being faced with some cryptic vocabulary – repository ... lower CASE ... reverse engineering are only a few examples. The following selection of terms and their explanations does not claim to explain all there is to know about CASE, but we hope that it will help anyone who wants to discover more about the subject to find their way through the jungle of jargon that surrounds it.

CASE (Computer Aided Software Engineering). This, in its widest sense, can include any piece of software that assists in the system development process. However, a CASE tool is usually a highly sophisticated collection of software, which can help developers in their work, especially in the areas of diagrams, cross-checking and documentation. CASE tools aim to automate the whole development process, from the initial identification of customer requirements to automatic code generation. Several CASE tools may be necessary to cover all aspects of a particular development.

Workbench. This includes all the tools necessary for a particular task. An analysis workbench, for example, may offer tools such as automated data flow diagrams, entity-relationship models, process definitions, labelling and definition facilities, and different types of report on the models produced. All tools in a particular workbench are integrated and designed to work together. There is also integration between different workbenches in the same CASE tool.

Environment. A set of integrated workbenches that cover most or all of the software development life cycle.

Repository. This is also referred to as the encyclopaedia or dictionary. It is the core of any CASE tool, holding all the information about the problem area and the computer system. The repository enables cross-referencing between diagrams, models and workbenches, to ensure overall consistency. It instantly reflects any changes in the developing system.

Upper CASE. Tools that support the analysis and design of software systems.

Lower CASE. Tools that support programming and testing.

Re-engineering and reverse engineering. Many organizations today depend, at least in part, on software systems that have evolved over a number of years but are still critical to the running of the business; these are known as *legacy systems*. Re-engineering and reverse engineering are methods in which the life of a legacy system may be extended. Sometimes the documentation of a piece of software is so sparse or confused that it is impossible to work out what the program was supposed to do in the first place. The process of reverse engineering feeds an old, inefficient piece of software into a CASE tool to find out its original purpose, and re-engineering restructures the code so that it runs more efficiently.

IPSE (Integrated Project Support Environment). These are CASE tools for the management of system development projects. They offer automated support facilities for project management, such as scheduling, cost estimating, version control, time accounting and problem recording.

Advantages of CASE tools

There are many advantages for a software developer in using CASE tools. The tools are based on structured techniques, such as entity-relationship modelling, or on object-oriented techniques, such as class and use case diagrams, which have been tried and tested by countless system developers. CASE does not involve a reversal of former approaches to the development process; the tools simply offer a faster, more efficient and less tedious way of doing the job. Many of the system developer's most time-consuming tasks can be carried out by the CASE tool repository, which can perform extensive cross-checking that would be too time-consuming to do by hand. This leads to fast, accurate detection of errors, omissions and inconsistencies.

Communication with the client during the development process can be greatly facilitated by using a CASE tool. CASE tools make comprehensive use of graphics and allow diagrams to be produced and altered with ease. This means that, when errors are discovered or when the client is not happy with a diagram, it is a relatively simple task to make the required changes. CASE tools facilitate and encourage reworking of models to ensure that the final design is exactly what the client wants.

The production of full, consistent documentation using a CASE tool is very much easier and more accurate than documenting a system developed by hand, since every item of data about the system is held in the central repository of the CASE tool. This is especially important on systems that take several man-years to develop, since it is likely that many different people will be working on the system during its lifetime. The extensive documentation produced and the speed of reworking mean that systems developed with CASE are more easily maintainable than systems developed by hand.

For developers who use prototypes as the basis of their work, all the advantages of prototyping, such as rapid development and improved communication with clients, are enhanced by using a CASE tool to build the prototypes.

On the management side, the advent of IPSEs (see the definition in the previous section) means that the technical details of developing a system can be totally integrated with the management of the development project. Project management is discussed in Chapter 10.

Some criticisms of CASE

Any tool or technique may give rise to criticisms and CASE is no exception. CASE tools are very expensive and represent a major investment for a software development company. Since many of the tools are tied to a particular development method, this involves a strong commitment to that method on the part of the company. This commitment, together with the size of investment, means that a development company may become fixed in a particular way of developing systems, rather than moving ahead to new techniques and approaches. A CASE tool involves further expense for the company in terms of time, effort and money since it needs to be maintained in the same way as any other piece of software. Moreover, there is a learning curve associated with the tools and specialized training is often needed. A software company would have to take this into account when considering whether it is worth investing in CASE.

The CASE tool can support, but must not replace, the work of the system developer. Development of software systems requires skills and talents, such as creativity and good communication, which cannot be performed by an automated tool. CASE is no guarantee of quality. What is beyond doubt is that these tools do save time and money in the system development process, and allow systems to be developed that would otherwise be unfeasible. However, the result of this should be that such savings will not be swallowed up by software development firms in buying and maintaining the CASE tool, but will be passed on to clients in the form of higher-quality systems that are delivered on time and within budget.

11.2 Prototyping

Designing anything involves building at least one model of it. A model offers us an abstraction, a particular view, of the object that is to be constructed. Children experiment with models using materials such as toy bricks, sand and paint. If a house is to be built or extended, several different models are made; these can include the architect's plans, the structural engineer's drawings and even the customer's rough sketches of what the house should look like.

Designing systems, like designing houses, involves modelling. In Chapters 4, 5, 6 and 7 we have already seen several examples of widely used modelling techniques offering a diversity of views of the system that is to be built. These techniques include such varied models as data flow diagrams, which show the movement of data in the system, entity-relationship models, which illustrate relationships between data objects, and state diagrams, which show how data is affected by events external to the system. Each of the techniques discussed shows a different aspect of the developing system, yet they all have one thing in common: a model created in one of these ways will always be just a model. However much care, time and effort the developer puts into the model, the model will never itself become a working system. CASE tools allow the developer to create versions of these models that can be manipulated and altered as the client requires, but even these dynamic models have to be implemented in code before the system is actually working.

Prototyping, on the other hand, is based on the concept of an implemented model. There are different ways of prototyping and different uses of prototypes, but what is common to all of them is that the system developer and even the client will be able to see and experiment with part of the system on the computer from a relatively early stage in the development process.

What is prototyping?

In common with many words in computer science, prototyping can mean very different things to different people. It is essential that a system developer, using prototyping with a team of other developers or when communicating with a client, makes sure that everyone concerned shares the same understanding of what the term 'prototyping' means. The following sections describe some of the most widely used prototypes.

Screen layouts. The most basic prototype consists of non-functioning mock-ups of sample screen and report layouts. The system developer will be able to demonstrate to the client exactly what will appear on the screen when entering data, for example, or what a typical management report will look like. The client can request changes to the designs and the models can be refined until the client is satisfied. With this sort of prototype the system developer will not, however, be able to demonstrate or experiment with the way the screens work or interact with each other. It is not possible actually to enter data or produce a report. The screens are simply design layouts and there is no working system behind them.

The disposable prototype. The power of modern programming languages makes it possible to build a system very much more quickly than with previous languages. With a disposable prototype, a model of the system (or part of it) is developed very quickly and refined in frequent discussions with the client. The model simulates the functionality of the proposed system, so that the client can see what it will do, but it is not backed up by detailed structured design. It is a similar situation to a dress designer who makes up the pattern for an exclusive outfit in cheap material to see how it will look. Once the designer and client have tried out various ideas on this model, and are happy about the design of the final garment, the actual outfit is made up along the lines of the model, but in a fabric of a much higher quality. Prototypes like this, in dress design or systems development, are very helpful in capturing requirements at an early stage in the development process, particularly if the client is not sure exactly what their requirements are or where the problems lie. In the area of computer systems these prototypes are also useful models for designing and refining the user interface – what the system will look like to the people who are actually going to use it.

One of the most important things about this sort of prototype is that it is, sooner or later, discarded – for all but documentation purposes. It is very tempting for designer and client alike to allow the model to be developed as the basis for the final system, but this is a path to disaster. The essence of prototyping like this, to capture client requirements, is speed. There is no place here for the detailed analysis and design that are essential for robust, reliable and maintainable systems. Keeping a prototype that has been built in this way, and developing it to become the final system, is as inappropriate as a dress designer selling the model made up in cheap material as an exclusive outfit. If the system developer decides to build this sort of prototype, it is also essential to decide exactly when and how it will be set aside, and then ensure that this is done when the time comes.

The evolutionary prototype. This prototype differs from the others in that, from the outset, the aim of the process is for the model to develop into the final system. This means that the system will generally be developed more quickly than by other, more traditional, methods and that the cost should be lower. It is important that extensive and detailed analysis of the problem is carried out before prototyping begins, otherwise the final system will be like a house built on sand. In this sort of prototyping, modern automated tools really come into their own; most developers will not attempt to build an evolutionary prototype without a CASE tool that can produce dynamic models and automatically generate code.

Prototyping and the client
With prototyping, the relationship between the developer and the client is much more that of a partnership than one of expert and customer. In the following sections we discuss the advantages and disadvantages of prototyping from the client's point of view, and explain what the client will be expected to contribute to the development process.

The advantages of prototyping for the client
The most obvious difference between prototyping and other techniques is that the prototype is an implemented model. When the client evaluates what is demonstrated, it is a live system, not a paper-based image of it. The client can have hands-on experience of the system during the development process and will be able to react to the design of the system before it becomes fixed. The client will find that seeing and experimenting with a dynamic model of the system makes it much easier to sort out exactly what the system should do and how it should look.

Effective prototyping will involve the client not merely in discussions with the system developers but also in realistic simulations of work situations. It will be possible to see how the system will perform in relation to tasks in the organization. If, for example, a particular report is produced at the end of each month, the client would be able to see from the prototype how the data for the report would be collected, entered, selected, ordered, formatted and finally printed. If the layout of the final report is not exactly what is wanted, it can be changed with relatively little effort. Examining and playing with a prototype is thus a very good way for the client to get to know the system and to judge its strengths and weaknesses in different situations.

The client's involvement in the system development process has many positive effects. It enables much easier communication with the developer about ideas, opinions and any changes that are to be made to the system. The client will also find that it is much easier to understand what the developer says about the system while sitting in front of a screen, rather than trying to understand documents and diagrams. The client's continuous involvement in the system throughout its development means that much less training is needed on how to use it when it is finally delivered. This, of course, applies equally to any employees of the client organization who have been involved in the development process. In this way considerable savings can be made on user training costs.

Even if the developer decides that a particular system is not suitable for prototyping, and that it will be developed in a more traditional way, it is still likely that prototyping will be used in the design of the user interface. This is the area that, more than any, benefits from a prototyping approach, since it is this that determines how the system will appear to the users. Details such as how the system will respond to a password (perhaps by a row of asterisks, a row of dashes or simply a message welcoming the user to the system) or the most effective design of a screen used in data entry, are the sorts of problems for which decisions are almost impossible to make effectively unless the client can actually see what will appear on the screen. At the start of the development the client may, for example, be quite set on a blue background with yellow text for all data-entry screens. It is only by seeing such a screen on the computer that the client can decide whether the colour scheme is really what is wanted.

The design and refinement of the user interface is one of the most important aspects of almost all information systems. No matter how good the system is in other respects, if it is not user-friendly no one will want to use it. In large organizations, it is obviously not possible for everyone who is going to use the system to express their opinion on how it will look, but it is important that as many people as is practicable, and certainly the principal users, have the opportunity to see what they will be working with.

The demands of prototyping on the client

For clients and users who are keen and able to play a part in the development of the system, the advantages of the prototyping approach far outweigh the disadvantages. The users' involvement in the development process brings an awareness of what is going on and means less frustration with apparent delays or lack of progress. When the developing system seems to be deviating from what the client asked for, the problem can be pointed out to the developer and together they can discuss the solution or necessary compromises. The users' attitude to the system tends to be more positive when prototyping is used as the development method, since involvement from the earliest stages brings a strong feeling of ownership of the growing system.

While this is an excellent way of developing a system, it does demand considerable input from the client and users in terms of time and effort. It is this degree of involvement that can be a real disadvantage of prototyping, from the client's point of view.

For successful prototyping it is essential that the client understands from the start just how much time and effort will be required. In large organizations this means that staff at all levels may be needed to contribute to the development process. Staff must be available to help define and, if necessary, repeatedly redefine models of the emerging system, screen and report layouts, and the various functions that the system is to perform; this can be very time-consuming and disruptive to the client's business.

However, there is no doubt that with a client and users who are able to contribute fully to the development process, the technique of prototyping offers the possibility of systems that are developed faster, more cheaply and that are much more likely to satisfy requirements. A project such as Just a Line, for example, would benefit greatly from prototyping, either in the early stages for capturing requirements or to create an evolutionary model that will eventually develop into the final system.

Prototyping and the system developer

Since there are several different views of prototyping, and since every problem is unique, it is difficult to generalize about what happens when prototyping is used to develop a system. The following section describes one way in which part of the Just a Line system might be prototyped. We then discuss the advantages and disadvantages of prototyping from the system developer's point of view.

Prototyping Just a Line

The use of prototyping does not mean that the developer can dispense with the traditional tasks of early systems development. The first step in a project such as Just a Line would be to carry out a feasibility study and decide whether it is really worth developing a new software system for the company.

It is also important to establish early on that the Just a Line problem is suitable for prototyping. From the interview notes we know that Just a Line is a small company with fairly standard problems of disorganized order processing, stock control, accounting and almost no marketing strategy. We can see how the overall system could be broken down into these four areas, which could each be prototyped separately, and then integrated to form the complete Just a Line system. We can also tell from the interview that Harry and Sue are very keen to modernize the running of their business and seem willing to help in any way they can. What is not clear is the amount of time either or both of them can spare to help with the prototyping process. This is something the developer would have to find out. It would also have to be established whether Harry and Sue are agreeable to this method of proceeding, and whether they understand and accept the amount of involvement required on their part.

In dealing with a larger and more complex system, the developer would now probably investigate the range of software tools available and decide which is appropriate for the job in hand. For a system the size of Just a Line, however, the use of sophisticated, expensive tools is not justified.

Having decided how to proceed and having drawn up an initial project plan (as described in Chapter 10), the system developer must now investigate the problem more fully and identify the various parts of the system that are to be developed. There are many ways in which this can be done; the method chosen will probably depend on the developer's own past experience and preferences. One possible way would be to develop a high-level model of the problem area with Harry and Sue, using some of the structured techniques described in Chapters 4, 5, 6 and 7. Whatever method is chosen, the developer should have a sound idea of both the data and the principal functions of the system.

In the case of Just a Line, we can identify the separate areas of marketing, order processing, stock control and accounting. The developer now takes one of these areas – marketing, for example – and examines it in greater depth. In order to get a clearer idea of what exactly the clients want in terms of, for example, a mail shot to regular customers, the developer may start to experiment with various ideas on the computer and invite Harry and Sue to comment and suggest improvements. Some CASE tools allow the system developer to design screens using a screen painter, which automatically generates code from the screens.

Figure 11.1 shows an early prototype screen for the Just a Line file of regular customers. The screen would be used by Harry and Sue when adding new customers to their file for checking customer details. At present the screen simply shows the name, address and phone number for each customer, together with a Just a Line customer code. The developer will now show the screen to Harry and Sue who can then express their opinions and make suggestions.

The developer will try to incorporate all Harry and Sue's suggestions into the screen design and will then demonstrate what has been done, to invite more comments. Figure 11.2 shows a later version of the screen, with additional fields to allow more customer details to be kept on file. In this version of the screen, the customer code has been altered to customer number and has been moved to a more prominent position, since this is how the Just a Line computer system will uniquely identify each customer. The address is more clearly displayed and an extra field has been added to show the date of the customer's last order. This new screen will now be shown to Harry and Sue. The cycle of demonstration and refinement will continue until both developer and client are satisfied with what has been produced.

```
Just a Line customer file

Name _____

Address _____

Phone No _____

Customer Code _____
```

Figure 11.1 An early prototype screen for the Just a Line customer file

```
                    Just a Line
                   Customer File

        Customer Number _____

        Name              _____

        Address           _____

                          _____

        Phone No          _____

        Date of last order _____
```

Figure 11.2 A later version of the prototype screen for the Just a Line customer file

Throughout this whole process Harry and Sue will be consulted regularly and asked to experiment with the developing system. The developer may set up special tasks for them to carry out, which make it possible to see how effective the system is. The developer will want to measure their performance in terms of speed, accuracy, and the number and types of mistakes made, as well as establishing how user-friendly they find the system to work with. Harry and Sue will be an integral part of the development process. They must be prepared to make decisions about their system and to take responsibility for them. Prototyping is hard work for everyone concerned, but it is also rewarding and generally results in a system that both developer and users can be proud of.

The advantages of prototyping for the system developer

Many of the advantages of prototyping for the system developer result from the direct involvement of the client, and include better communication and more meaningful feedback on the way the system is developing.

The tasks of capturing the nature of the current problem and the client's future requirements are much less prone to error and misunderstanding when using a live prototype (rather than static paper models). This applies also to prototypes of the user interface, which allow the developer to show the client from an early stage exactly what the system will look like to its users.

Quite apart from the improved client communication, working with a live dynamic model offers the system developer other advantages. Static models, fixed in the stages of the traditional system life cycle, are at best cumbersome and at worst impossible to change. Yet in developing systems change is the norm, whether because of errors discovered, outside factors such as changes in tax rates or simply the client changing his or her mind. The prototype model is built to accommodate change; the developer expects to have to go round the loop of demonstrating to the client and refining the model several times before agreement is reached.

The inherent flexibility of the prototype model allows the designer to evaluate design decisions before they become fixed. Progress of the system in relation to the client's wishes can be monitored and alterations can be made accordingly. By setting up realistic simulations, such as a typical mail shot, the developer can judge how the system will perform and identify where improvements can be made.

Whatever precise interpretation a system developer gives to the term prototype, the basis of the approach is the live dynamic model. It is the model, together with the increased client involvement that it encourages, that gives prototyping its advantages as a method of developing software systems.

The disadvantages of prototyping

So far, we have emphasized all the advantages of prototyping. We have mentioned the extra time and effort that has to be put in by the user, but it may appear that the process is all plain sailing for the system developer. Unfortunately, this is not the case.

Prototyping is essentially an iterative process. The developer has to realize that it may not be right first, or even fifth, time round. This constant revising and refining is often difficult to accept for a more traditional software system developer who has been accustomed to using a structured development method. Many developers understandably feel protective about diagrams or code that may have taken a lot of time and effort to write and that may now have to be scrapped.

Not only is the methodological way of viewing system development inappropriate for prototyping, but the traditional job definitions no longer apply either. The systems life cycle demanded the participation of requirements engineers, systems designers and programmers – all of whom were specialists in their own particular area, but did not need to know in detail what the others were doing. A developer who is prototyping all or part of a system not only needs to be adept at capturing requirements and designing software, but must also be able to program in at least one modern programming language. The traditional analyst–programmer split no longer applies, and a developer who cannot follow a system through from the initial idea to the final implementation will be of little use in prototyping a system.

Effective prototyping demands not only people who have been trained in a new way but also more powerful software tools than the traditional development methods. Prototyping needs speedy implementation and flexibility of software produced. For this, modern programming languages are essential and, for systems of any size, CASE tools will also be needed.

Whereas communication with the client is much easier using prototyping, the amount of communication needed between different software developers working on a project can be much more difficult. If a system is split up into different areas for separate development of prototypes, it is essential that these separate parts can all eventually be integrated into one single, cohesive system. This involves a large overhead of communication between developers prototyping the separate parts. There can come a point, with a large and complex system, where it needs to be split into so many separate parts that the communication overhead makes prototyping impractical as a development method. The nature of prototyping means that it is really a one- or two-person job. It will always be useful in developing certain parts of systems, such as the user interface, but it has not yet been proven as a suitable method for developing all types of large and complex systems.

This problem of communication between developers is aggravated by the fact that prototyping is not a structured method and therefore has no underlying framework to aid management of the system development project. Traditional development methods divide the development process into distinct stages, and prescribe reviews and walk-throughs at specific points. The technique of prototyping does not have these in-built milestones. New working guidelines have to be devised around the activities and deliverables of prototyping and, unless it is effectively managed, the development of a system using prototyping can easily get out of control.

Although prototyping does have drawbacks, both for the developer and the client, there is no doubt that with the right people, in the right circumstances, it is a fast and effective way of developing high-quality systems, and can encourage a friendly, productive relationship between the software developer, client and users. In cases where it is not appropriate to use prototyping as the main development approach, it is still useful for the developer to prototype those areas of the system, such as the interface, where user involvement is essential.

11.3 Rapid Application Development

Rapid Application Development (RAD) was originally developed by James Martin (1991) and is one of the most widely used and successful implementations of prototyping. RAD encompasses a strategy for system development that relies on a high degree of user participation and exploits modern high-powered software tools. The aim of RAD is to shorten the life cycle and to produce information systems more quickly in order to respond to rapidly changing business requirements, such as those of e-commerce and other web-based systems. Figure 11.3 illustrates the difference between RAD and a typical traditional system life cycle.

A typical RAD development consists of four phases, all of which involve the participation of users, particularly those who are experienced in RAD. The phases are requirements planning, user design, construction and cutover. Generally, the highest level of user involvement occurs during the requirements planning phase, where the focus is on solving business problems and ascertaining information requirements. During the requirements planning phase, developers seek to obtain input from users from all levels of the client organization, in order to determine an agreed set of system requirements. The user design phase is based on a prototyping cycle, involving experienced users and developers. This phase often consists of a series of workshops, each of which may last up to three days. During a workshop, CASE tools are used to build prototypes, which undergo user evaluation, subsequent refinement and further evaluation, until both users and developers are satisfied. During the construction phase, code is generated from the CASE prototypes and the new system is validated by users. In small RAD developments, user design and construction are often treated as a single phase. The final phase, cutover, covers system testing, user training and introduction of the system into the client organization.

Using RAD as a development strategy has proved to be a successful approach for many software companies. RAD has a number of obvious advantages:

- the high degree of user participation in the whole process means that resistance to change in the organization is minimized and the new system is welcomed
- systems are undoubtedly developed and delivered more quickly than with traditional development approaches
- the speed of development and the use of relatively small teams mean that RAD projects tend to be cheaper than their traditional counterparts
- the speed of development and the close involvement of users also mean that RAD systems are more closely related to the current needs of the business.

RAD, however, is not a panacea and inevitably it has some drawbacks:

- RAD works only for certain types of system – business information systems that need to respond quickly to changing requirements; there are many other types of system (safety and security-critical, for example) for which it is simply not appropriate

RAD life cycle	Traditional life cycle
Requirements planning	Problem definition
User design	Flexibility study
Construction	Requirements engineering
Cutover	Design
	Implementation
	Maintenance

Figure 11.3 RAD and traditional system life cycles

- the emphasis on speed of development means that sound software engineering practices are sometimes given low priority or omitted altogether; this can result in problems with consistency, programming standards, documentation and system administration issues, such as backup and database maintenance
- although RAD claims to be a relatively cheap way of developing systems, the heavy reliance on user commitment and participation can push up costs for the client organization, since employees may be taken away from their usual tasks
- although changes affecting the system design can be implemented easily during development, long-term modification of the system is more problematic because of the relative lack of standard documentation.

RAD is an increasingly popular approach to systems development, principally because of the growth in web-based and e-commerce systems, which demand fast development methods that can respond to rapidly changing requirements. For systems like these, RAD is a suitable and effective development method. Developers should be aware, however, that the benefits of RAD are counterbalanced by a number of risks, and that, for many systems, this approach to development is neither feasible nor appropriate.

11.4 Agile Methods

RAD has been around for a number of years and is still widely used, particularly in the development of business information systems. Increasingly, however, developers are turning to new methods that are strongly influenced by prototyping and RAD, but that are more clearly defined, introducing their own features and procedures; these are known as *Agile Methods*.

A typical Agile development begins with project initiation; this establishes the high-level requirements and the justification for the project. In many ways project initiation is similar to the feasibility study in a traditional system development project. The next stage is project planning. The plan is based on a number of releases, each generally about three months long. A release is made up of timeboxes, which are generally three to four weeks long, with a definite date for completion of the timebox and delivery of the product. Each timebox works on a defined sub-set of the system requirements and has its own project plan.

There are a number of different Agile Methods, including DSDM (Dynamic Systems Development Method) which is probably the closest to the original RAD approach, and XP (eXtreme Programming). XP has become well known particularly because of its use of *pair programming*. This involves two programmers working together, sharing the mouse, keyboard and monitor, collaborating on the same design, code and testing. XP claims that pair programming encourages developers to talk through their ideas and the problems they encounter, resulting in faster development and a better-quality product.

Although each Agile Method has its own special development features, they are all based on the same fundamental principles, as described below.

- Agile Methods are based around iterative, rather than structured development, with continuous delivery of software from an early stage in the project. Frequent releases allow adaptability and the ability to respond easily and quickly to changing requirements and situations.

- The focus is always on the working product, rather than on mountains of documentation, although the number of documents that have to be produced varies widely between methods, with DSDM producing a number of documents, but XP relatively few.
- There is a strong focus on customer involvement. The customer is viewed as an integral part of the development team, with responsibility for prioritizing the requirements and selecting those that will be developed in the next release.
- Selection of the right team is seen as the key to the success of the development project. Agile team members have to have the necessary experience, expertise and skills to be able to carry out tasks at all stages of the development process. In addition to technical skills, it is essential that they are good communicators.

11.5 Open source development

Open source is a new way of organizing the development of software and disseminating the source code. It is becoming increasingly popular, with highly successful applications such as the Linux operating system and the Apache web server.

With most software, users have no access to the source code, so it is very difficult, if not impossible, to understand exactly how it works or to make changes. With open source software (OSS) the code is on the Internet and freely available to all users, who can modify it, add to it and redistribute it as they please, subject to certain conditions.

The rights that come with open source software are:

- the right of access to the source code
- the right to make copies of it
- the right to modify it
- the right to distribute it.

These mean that development of OSS is frequently carried out by a number of individual developers working together or separately on the same program.

Open source allows constant peer review, thus ensuring a high level of quality in the most widely used programs. On the other hand, there are large numbers of open source projects on the Internet where the programs are unfinished and of mediocre quality. Open source does not in itself guarantee quality software, but it does bring far-reaching changes to the way in which software is developed and distributed.

11.6 Object-oriented development

Software systems developed over the last 30 years using the structured methods described in this book have shown significant improvements in terms of robustness, reliability and maintainability. However, as the problems tackled by software developers have become progressively larger and more complex, it has become apparent that structured methods are not a complete answer. Despite all the effort that has gone into improving the software development process, it is still relatively rare for a project to be delivered on time, within budget and without serious faults. Software practitioners want a new approach to system development that will produce software that is more maintainable, testable, reusable, and able to cope with large and complex systems. The object-oriented (O-O) approach, based on the data objects in a system rather than on its functionality, claims to deliver these qualities.

Background to object-oriented development

Concepts from which the object-oriented approach is derived have been around since the 1960s. The programming language Simula was developed to write computer simulations of real processes. Program units in Simula were not based on functionality but on the real-world objects to be simulated in the software. Ideas generated by the Simula experiment were taken up and developed by the team that produced Smalltalk in the Xerox Palo Alto Research Center. There are now many object-oriented programming languages, including Smalltalk, C + + , Java and Eiffel. The initial interest in object-orientation focused on programming language issues. More recently, however, O-O ideas have been applied to the whole software development process – analysis, design and implementation. New modelling techniques, the equivalent of techniques used in structured modelling, have been developed to model the application domain in object-oriented terms.

Objects and the class diagram

The basic building block of object-oriented software is the **object**. Software objects are derived from and model the real-world objects in the application domain. Whatever type of software system you are developing, whatever problem you are trying to solve, it will feature certain entities, objects or things. These objects or things form the subject matter of the system. A mail order system will feature objects such as customers, orders and products; a library system will feature objects such as members, books, loans and reservations; a restaurant booking system will have customers, tables, reservations and cancellations; a system to simulate traffic behaviour might have traffic lights, cars, bicycles and pedestrians. All of these objects are understandable features of their problem domain and will have a software representation in the system being written. Objects in the real-world therefore translate into objects in the software system.

The O-O model used to represent the objects in a system is the class diagram. Every object belongs to a **class** that defines its structure: the number and nature of its attributes and **operations** (see below). For example, in the Just a Line system, Customer might be a class with attributes customerNo, name and address. The class diagram has much in common with the E-R model (see Chapter 6): classes in many respects resemble entities (or entity types); objects resemble occurrences of entities. The class diagram also models relationships between classes, some of which, superficially, have much in common with entity relationships. If the Just a Line system were redesigned using an O-O approach, its class diagram might look like the one in shown in Figure 11.4.[1]

Classes are represented by rectangles with the name of the class in the top section, the attributes of the class in the middle section and the class operations in the bottom section. The lines that connect the classes are known as **associations** and indicate that these classes have some sort of relationship. The numbers and asterisks indicate the **multiplicity** of the relationship. The asterisk means zero, one or many. In Figure 11.4 the multiplicity symbols tell us that:

- a customer may have any number of orders between zero and many (number unspecified)
- an order is for only one customer

---

[1] For the purposes of describing object-orientation we have assumed a version of the Just a Line system that checks and decreases stock levels.

- an order may have one to many order lines
- an order line will belong specifically to one order
- an order line will refer to only one product
- a product may appear on zero or many different order lines
- a product will be supplied by only one supplier
- a supplier may supply one or many products.

Alert readers will notice that the classes in Figure 11.4 correspond exactly to the entities in the E-R model of the Just a Line system in Figure 6.26. However, a class differs in one fundamental respect from an entity in that it has *behaviour*. The functionality of an O-O system is delivered by its operations, which are divided up between its classes. The diagram in Figure 11.4 shows a few of the operations that might be specified for the Just a Line system. What an object is able to do is specified by the operations on its class. At the very least, an object will be able to store, edit (set) and display (get) values for its attributes. For example, an object of the Product class will be able to store, edit and display its product number, description, packet size, price and stock level (see Figure 11.5). As this can get a bit long-winded the so-called set( ) and get( ) operations are frequently omitted from the class diagram, although they will be there in the code.

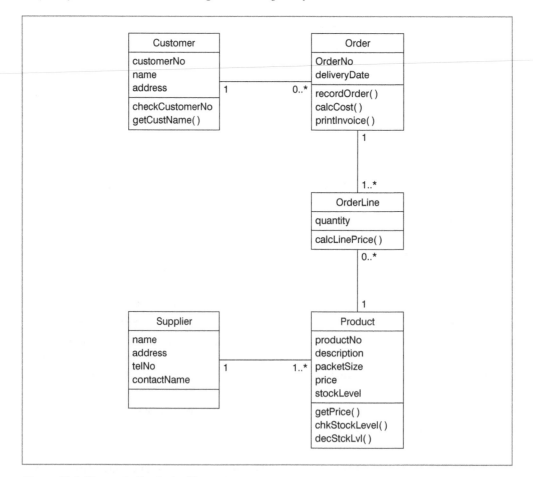

Figure 11.4 Classes in the Just a Line system

```
┌─────────────────────────────┐
│           Product           │
├─────────────────────────────┤
│ productNo: String           │
│ description: String         │
│ packetSize: Integer         │
│ price: Money                │
│ stockLevel                  │
├─────────────────────────────┤
│ setProductNo( )             │
│ editProductNo( )            │
│ getProductNo( )             │
│ setDescription( )           │
│ editDescription( )          │
└─────────────────────────────┘
```

Figure 11.5 The Product class

Every object in an O-O system has a name, knows which class it belongs to, has values for its set of attributes and knows it can *do* certain things. Figure 11.6 is an object diagram showing a Product object (called aProduct) with values for its attributes. Operations are not shown on the object diagram – they are only ever shown on the class diagram.

A system, however, will have to do more than store, edit and display attribute values. All of the system functionality is delivered by the operations defined on the classes; operations cannot exist apart from classes. Functionality is allocated to the class where it seems most appropriately to belong. For example, the class OrderLine (see Figure 11.4), representing one line in an order, in addition to having operations to store, edit and display the value of its attribute, quantity, might have an operation calcLinePrice( ) to calculate the price for an order line.

The relationships between classes are also significantly different from those on an E–R diagram. You may have noticed that on the class diagram in Figure 11.4 there is a total lack of foreign keys: there is no reference in OrderLine to Product, no reference in Order to Customer, nor in Product to Supplier. This is because objects are directly linked by *object identifiers*, not by foreign keys. Association relationships indicate that there must be a *navigable path* between objects of related classes by which they can communicate. In an object-oriented system, objects collaborate to achieve the required system functionality. An object will be sent a request to perform its part of the workload. In turn, it may send a request to another object to perform the next part of a task. The sending of such requests to achieve inter-object collaboration and communication is known as *message passing*. For example, for an object of the OrderLine class to perform its calcLinePrice( ) operation, it will have to multiply the relevant product price by the quantity ordered. The quantity ordered is held in its own attribute, quantity. However, the price is stored in the relevant Product object. The OrderLine object knows which Product object this is, because it knows its object identifier (otherwise known as

```
┌─────────────────────────────┐
│      aProduct: Product      │
├─────────────────────────────┤
│ productNo = SD00678         │
│ description = birthday card │
│ packetSize = 10             │
│ price = £1.78               │
│ stockLevel = 35             │
└─────────────────────────────┘
```

Figure 11.6 An object of the Product class

its reference or pointer); ultimately the OrderLine object knows the memory address of the relevant Product object. The OrderLine object sends a *message* to its Product object asking for the price. Once it knows this, it can do the calculation.

The data attributes in an O-O system are organized in such a way that they can be accessed only by using the operations defined for the class to which the attributes belong. In this way access to data is controlled – in general an object's data can be accessed only on receipt of a message requesting that it execute one of its operations. For example, if we need to change the price of a Product object, we must send a message to the relevant object asking it to execute its editPrice( ) operation. We cannot directly alter the price attribute. This way of protecting data is known as *data hiding*; packaging together data and operations is known as *data encapsulation*. Data hiding and data encapsulation are two of the distinguishing features of O-O systems. They make O-O systems more robust when it comes to maintaining the system, because data is less likely to be altered accidentally as the knock-on effect of a change elsewhere in the system.

O-O systems are characterized by messages being passed between objects, whereas structured systems, using functional decomposition, are characterized by data being passed between functions. Proponents of the object-oriented approach claim that this means that the data in an O-O system is better protected than that in a traditional structured system. O-O enthusiasts further claim that as the functionality required of a system (what it has to do) will change much more rapidly than the data in a system, it makes more sense to base the

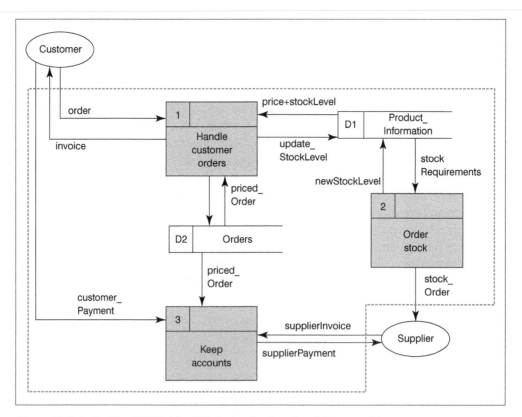

Figure 11.7 Just a Line CLDFD (simplified) showing functional decomposition

structure of a system on its data rather than its function. The contrast between these two styles of design can be seen by comparing Figures 11.4 and 11.7. In Figure 11.7 the three process boxes model the main functions in the system; the data in the data stores can be freely updated by any of the processes. The lines connecting the boxes represent data flowing round the system. In Figure 11.4 the main decomposition of the system is into classes of objects; the lines connecting them represent navigable paths along which messages can be sent.

Use case modelling

Although operations on classes ultimately deliver the required system functionality, a lot of work has to be done before this stage is reached. O-O system developers must do the same sort of investigation of user requirements as we described earlier in this book (see Chapter 3). However, to model user requirements they use a technique known as use case modelling. Figure 11.8 shows a simple *use case diagram* of the Just a Line system.

Each ellipse models a *use case*. A use case represents one of the system activities, a well-defined part of the system functionality seen from the point of view of the system users. Each use case depicts what the users of the system view as a separate task; for example, calculating the stock requirements, maintaining the product list, marketing, etc. The stick figures represent **actors**: users of the system. The lines linking the actors to the use cases tell us which parts of the system each actor uses; for example, both the Clerk and the Administrator can maintain the customer list, but only the Clerk will process orders. Actors

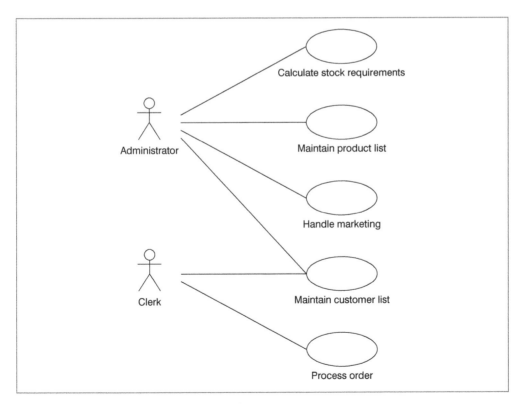

Figure 11.8 Use case diagram for the Just a Line system

don't necessarily represent actual people or job titles; they represent a role in relationship to the computer system – if Harry logs on as the Administrator he can maintain the product list, handle marketing etc.

Each use case has a goal, such as to successfully process an order, or to update the customer list. The detail of what happens in each use case is explained in an accompanying use case description and a set of **scenarios**. The use case description gives a generic description of what the use case does. It describes what usually happens, defines the use case goal and outlines the main alternative courses of action. Figure 11.9 is a use case description for the Process order use case. It is important to realize that all parts of the use case model – the use case diagram, the use case descriptions and the scenarios – see the system from the user's perspective. The nitty-gritty of what is happening behind the scenes, from the computer's perspective, is modelled elsewhere, primarily in the class and interaction diagrams.

Scenarios outline, step by step, a specific sequence of interactions between the actor and the system during the execution of a use case. For example: when Sue processed an order for an existing customer who ordered stock items all of which could be supplied from stock; when Sue processed an order for an existing customer and they had run out of Last Waltz anniversary cards; or when Sue processed an order for a new customer that had two items that were not lines that had ever been stocked by Just a Line (see Figures 11.10, 11.11 and 11.12).

Scenarios are a useful way of describing clearly and in detail what happens in complicated situations. At the requirements engineering stage, system developers find them a very useful tool for teasing out how an existing system works. They can also be a good way of helping the user to specify exactly how the new system is to work. As they spell out what happens normally and all the main variations from the norm, they come in very handy at the testing

Use case:	Process order
Actor:	Clerk
Goal:	To process an order

Overview:

Orders from existing customers come in by post. The clerk checks the customer details on the computer then inputs the new order details. The computer checks the stock levels, prices the order and prints an invoice.

Typical course of events:

Actor (Clerk) action	System response
1. Receives customer order	
2. Keys in customer name	3. Displays the customer details
4. Keys in details of order	5. Checks and prices order
	6. Prints invoice
7. Makes up order	
8. Posts order with invoice	

Alternative courses:

Steps 2 and 3   The order is from a new customer so the Clerk has to key in the customer details.

Steps 4, 5 and 6 If customer orders an out-of-stock or unstocked item, that item is not priced and does not appear on the invoice.

Figure 11.9 Use case description of the Process order use case

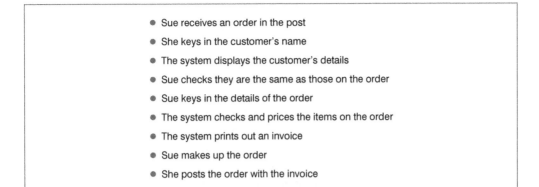

Figure 11.10 Successful scenario for the Process order use case

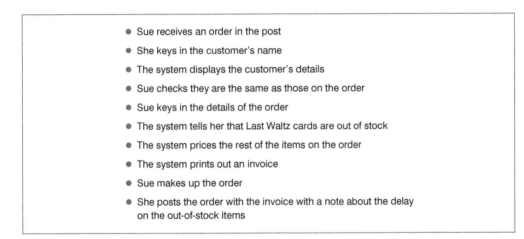

Figure 11.11 Process order scenario for an out-of-stock item

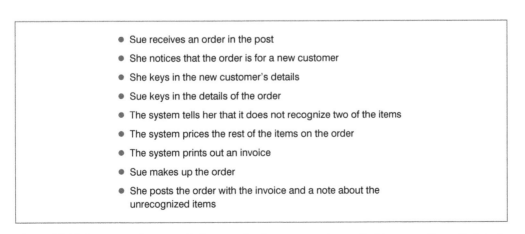

Figure 11.12 Process order scenario for an order for a new customer that has some items that Just a Line doesn't stock

stage too. However, to describe every possible scenario for every use case can get very long-winded, so developers tend to use them sparingly – all scenarios should, in any case, be covered by the use case description.

Interaction diagrams

Interaction diagrams model the inter-object messaging that happens during the execution of a particular use case scenario. While a scenario models what happens in a particular instance of a use case from the user's perspective, an interaction diagram models what happens on the other side of the computer screen, from the computer's perspective. It shows how a set of interactions between user and the system translate into messages between objects. There are two types of interaction diagram: *sequence diagrams* and *collaboration diagrams*. In this short introduction to object-orientation we will concentrate on sequence diagrams.

Figure 11.13 shows a sequence diagram that models a successful execution of the Process

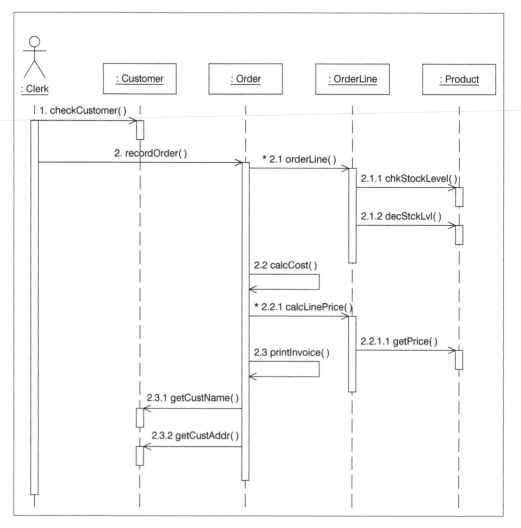

Figure 11.13 Sequence diagram for a successful scenario of the Process order use case

order use case. The objects involved in the delivery of the use case goal (<u>: Customer</u>,[2] <u>: Order</u>, <u>: OrderLine</u>, <u>: Product</u>) are lined up along the top of the diagram; the actor involved (<u>: Clerk</u>) appears on the top left-hand side of the diagram. The vertical lines below each object, called *lifelines*, represent the objects' existence during the execution of the use case. The thin rectangles on the lifeline indicate the points when an object is active. Messages are sent to the objects asking them to contribute their part in the functionality by executing an operation. The messages are modelled as labelled arrows. The diagram reads from top to bottom, specifying the order in which the messages are sent.

Message number one, checkCustomer( ), is sent from the Clerk (via some as yet unspecified menu object) to the Customer object that stores the details of the existing customer, asking it to display the customer details on the screen for the Clerk to check. Message number two creates a new Order object and starts the process of recording the order details. The next message is labelled 2.1, indicating that it forms part of the recordOrder( ) operation; orderLine( ) creates a new OrderLine object. This is done for each order line on the order – the repetition is indicated by the asterisk (*) before message number 2.1. The new OrderLine object then sends a message, chkStockLevel( ), to its Product object to check that there is enough stock to meet the order. If so, a decStckLvl( ) message is sent to update the stock level attribute.

Control then returns to the Order object, which is still in the process of executing its recordOrder( ) operation. It sends a self-call message – calcCost( ) – to one of its own operations. A self-call message is modelled by a message arrow that returns to the sending object's lifeline. The calcCost( ) operation asks each OrderLine to calculate its line price and these are accumulated by the Order object. The Order object then prints an invoice for the order; to do this it has to ask the Customer object for the appropriate name and address.

Interaction diagrams link the use case diagram to the class diagram. They show how objects (of the classes specified in the class diagram) send messages to each other requesting the execution of operations (specified on the classes) during the interaction involved in an instance of a use case. Interaction diagrams are very useful for specifying code and provide useful road maps through the code once it has been written.

There are several other modeling techniques used in O-O development. The three most popular are CRC cards, state diagrams and activity diagrams. **CRC (class–responsibility– collaboration) cards** are used to help in the allocation of functionality to classes. Each class is represented by a card. The system developers then do a walk-through of the main use case scenarios, allocating functionality to classes as they proceed. The attributes and functionality allocated to a class are called its **responsibilities**. For example, the OrderLine class is responsible for knowing the quantity ordered and for calculating the line price. At a later stage the functionality given to a class is turned into a set of attributes and operations on that class. CRC cards are also used to explore collaborations: whether a class can fulfil its responsibilities on its own or whether it needs to collaborate with another class to do so. We saw earlier, for example, that the OrderLine class needs to collaborate with the Product class to fulfil its responsibility to calculate the line price.

State diagrams in object-orientation work in the same way as in structured systems modelling (see Chapter 7) except that they model what can happen to a class of objects rather than to an entity.

---

[2] <u>: Customer</u> means an object of the Customer class.

Activity diagrams are used to model some processes in detail. They are typically used to specify what happens in a complicated operation. They can also be used to model the working of a use case, as an alternative to a use case description. Activity diagrams use a flow-chart technique, concentrating on showing flow of control. They are useful for modelling alternative courses of action and for modelling activities that can take place in parallel.

Advantages of the object-oriented approach

The object-oriented community claims that an O-O system is easier to maintain than one based on functional decomposition. Objects resemble the real-life object they represent; this means they behave predictably, in ways that are more intuitive and easier to understand. Code that is easy to understand is easier to maintain. Classes aim to be relatively small in size, which means that a maintaining programmer has less to hold in his head at any one time; this also contributes to easier maintenance. The code is also allegedly easier to follow because classes aim to be self-sufficient: operations and data that are required to fulfil its responsibilities are contained within the class. Where one class needs to collaborate with another, this is clearly indicated by the inter-object messages. Self-contained classes suffer less from the ripple effect associated with the introduction of modifications to code. The data hiding achieved by encapsulation also contributes to containing the effects of modifications. The object-oriented community also claims that an O-O system is more stable than one based on functional decomposition. O-O code is structured into classes; this makes the code more robust than code structured by functionality, as the data in a system changes less frequently than the functionality.

The size, *cohesion* and self-sufficiency of classes makes them easier to store in *class libraries* and to *reuse*; in addition, it makes them easier to test.

Summary

With the increasing size and diversity of system development projects, CASE tools have become an essential aid for software development teams. CASE tools provide automation of virtually all the tasks that take place during development, including drawing diagrams, checking for consistency and producing code. The tools allow many systems to be built that would otherwise not be possible.

Today's business systems typically have rapidly changing requirements and users who are familiar with many aspects of computer technology. In this type of situation, prototyping, Rapid Application Development and Agile methods provide an effective and responsive approach. In addition, the availability of open source software on the Internet has led to new ways of developing and disseminating programs.

The object-oriented approach offers many improvements, in terms of development method, techniques and notations, over traditional structured development. Many of the problems identified as those causing software to be insufficiently robust and hard to maintain are specifically addressed in the O-O approach. These improvements are claimed to be across the whole software development process – requirements capture, analysis, design and code.

Exercises and topics for discussion
What can you remember?

You will find all the answers in the chapter

a) List three types of automated tool that would be included in the general definition of CASE.

b) What are the advantages of automated diagrams for the client?

c) What is a repository in a CASE tool?

d) What is meant by a CASE environment?

e) What is a legacy system?

f) What is the purpose of an IPSE?

g) What is the essential difference between a prototype and a model such as a DFD or an E-R model?

h) What is a disposable prototype?

i) What is an evolutionary prototype?

j) List two advantages of prototyping for the client and for the system developer.

k) List two disadvantages of prototyping for the client and for the system developer.

l) How does RAD differ from a traditional system life cycle?

m) What are the main resources needed for RAD development?

n) List two advantages and two disadvantages of RAD as a development approach.

o) In Agile Methods, what does the project plan generally consist of?

p) Name two Agile Methods.

q) How is the customer viewed in an Agile development?

r) List four rights that users of open source software are entitled to.

s) Name two successful open source applications.

t) What is the difference between an object and a class?

u) What is the main difference between an entity and a class?

v) Why do classes not have foreign keys?

w) Explain the terms data hiding and data encapsulation.

x) What is the main difference between a system designed using functional decomposition and an O-O system?

y) What is modelled by a use case?

z) What is modelled by an interaction diagram?

11.1 If you have access to a CASE tool, copy one of the diagrams from the Just a Line system in Chapter 4.

11.2 'CASE tools will never really solve the problems of the software developer, since they address the symptoms rather than the causes.' Discuss in a group what is meant by this statement and to what extent you agree with it.

11.3 Design a prototype screen that will be used to enter details of an order for a Just a Line customer and calculate the amount owing. Using someone else in the group as the client, demonstrate and refine your customer order screen.

11.4 Design a report for Harry and Sue that will allow them to see how well various lines have sold during the past year. The report should be detailed enough to show seasonal variations in demand. Discuss and refine the report layout with another member of the group.

11.5 For which of the following systems do you think RAD would be a suitable development approach?

- A system for monitoring a patient's heart rate in intensive care.
- A system to publicize jazz gigs on the web.
- A web system to enable a surgeon to operate on a patient who is in a different location.
- An Internet banking system.
- A system for a high-fashion clothes company.

11.6 With a friend, as pair programmers, write a small program – for example, to work out the number of words in a sentence. Make a note of any ways in which this experience differs from writing a program on your own.

11.7 Look at the website http://sourceforge.net/ and find out more about open source software.

11.8 Based on your understanding of the Just a Line system, write a use case description for the use case Maintain product list.

11.9 Write a scenario for the Maintain product list use case for a situation where a new batch of cards arrives and the stock level needs to be updated.

11.10 Draw a use case diagram for the automatic ticket machine case study described in Exercise 4.3.

References and further reading

Bennett, S., McRobb, S. and Farmer, R. (2005) *Object-Oriented Systems Analysis and Design using UML*, 3rd edn, McGraw-Hill, London.

Britton, C. and Doake, J. (2000) *Object-Oriented Systems Development: A Gentle Introduction*, McGraw-Hill, London.

Britton, C. and Doake, J. (2004) *A Student Guide to Object-Oriented Development*, Elsevier Butterworth-Heinemann, Oxford.

Fowler, M. with Scott, K. (2000) *UML Distilled: A Brief Guide to the Standard Object Modeling Language*, 2nd edn, Addison-Wesley, Reading, MA.

Martin, J. (1991) *Rapid Application Development*, Macmillan, New York.

Pfleeger, S.L. (2001) *Software Engineering, Theory and Practice*, 2nd edn, Prentice-Hall, Upper Saddle River, NJ.

Pressman, R.S. (2000) *Software Engineering: A Practitioner's Approach*, 5th edn, European adaptation by Darrel Ince, McGraw-Hill, London.

Sommerville, I. (2004) *Software Engineering*, 7th edn, Addison-Wesley, Wokingham.

APPENDIX A: JAYS NEWSAGENTS CASE STUDY

1. Case study description

Mrs James of Jays Newsagents has asked you to investigate the possibility of computerizing its newspaper delivery system. Jays is a small local newsagent in the village of Pinkington. It has about 250 customers.

Orders are recorded in a large customer book that contains an entry page for each customer showing their name and address with their magazine, local paper, periodical and newspaper requirements (see Figure A.1). The customer book also records the cost of the publications, the delivery charge and the total amount due each week. The rest of the page consists of duplicated, perforated slips (the top copy is pink, the bottom copy, white). There is one slip for each week. This book is in alphabetical order by customer surname.

The newsagent also has a round book (see Figure A.2); this is kept in round order. It records the street, the house number and the requirements for each day.

When a customer requests a temporary amendment to their regular order (known as an exception) the newsagent writes this in her page-a-day diary (see Figure A.3). This includes extras and cancellations. For example, a customer might want to have *The Times Educational Supplement* delivered for the next three weeks only. The usual exceptions, however, are cancellations because of holidays. If the amendment is permanent the newsagent alters the entries in both the customer book and the round book.

Every Friday evening the newsagent goes through the customer book and writes the amount due on the perforated slip for that week, taking into account any exceptions that have been recorded in her diary. When a customer comes in to pay the bill, the top (pink) copy of the relevant slip (or slips) is torn off and given to the customer as a receipt. The newsagent keeps the white copy.

To organize the delivery boys and girls, the newsagent goes through her round book, with her diary, writing customer addresses on the newspapers and magazines ordered. She puts the papers into the delivery bags.

Periodically the newsagent goes through the customer book and prepares bills for all customers who have more than four weeks' slips in the customer book. The bills are delivered with the papers.

ADDRESS *Church Lane*		NAME *Chaney*		NO. *4*

Mags & Locals	Periodicals	Sunday Papers	Daily Papers	Delivery Charge *1.10*
				Amt *6.20*
		IND SUN	*IND Mon - Fri*	
		1.80	*50*	
			IND Sat	
			80	

w/e Sat. April 26th '03	w/e Sat. Jan. 18th '03	w/e Sat. Oct. 12th '02	w/e Sat. July 6th '02
Extras 4	Extras 42	Extras 28	Extras 14
w/e Sat. April 19th '03	w/e Sat. Jan. 11th '03	w/e Sat. Oct. 5th '02	w/e Sat. June 29th '02
Extras 3	Extras 41	Extras 27	Extras 13
w/e Sat. April 12th '03	w/e Sat. Jan. 4th '03	w/e Sat. Sept. 28th '02	w/e Sat. June 22nd '02
Extras 2	Extras 40	Extras 26	Extras 12
w/e Sat. April 5th '03	w/e Sat. Dec. 28th '02	w/e Sat. Sept. 21st '02	w/e Sat. June 15th '02
Extras 1	Extras 39	Extras 25	Extras 11
w/e Sat. Mar. 29th '03	w/e Sat. Dec. 21st '02	w/e Sat. Sept. 14th '02	w/e Sat. June 8th '02
Extras 52	Extras 38	Extras 24	Extras 10
w/e Sat. Mar. 22nd '03	w/e Sat. Dec. 14th '02	w/e Sat. Sept. 7th '02	w/e Sat. June 1st '02
Extras 51	Extras 37	Extras 23	Extras 9
w/e Sat. Mar. 15th '03	w/e Sat. Dec. 7th '02	w/e Sat. Aug. 31st '02	w/e Sat. May 25th '02
Extras 50	Extras 36	Extras 22	Extras 8
w/e Sat. Mar. 8th '03	w/e Sat. Nov. 30th '02	w/e Sat. Aug. 24th '02	w/e Sat. May 18th '02
Extras 49	Extras 35	Extras 21	Extras 7
w/e Sat. Mar. 1st '03	w/e Sat. Nov. 23rd '02	w/e Sat. Aug. 17th '02	w/e Sat. May 11th '02
Extras 48	Extras 34	Extras 20	Extras 6
w/e Sat. Feb. 22nd '03	w/e Sat. Nov. 16th '02	w/e Sat. Aug. 10th '02	w/e Sat. May 4th '02
Extras 47	Extras 33	Extras 19	Extras 5
w/e Sat. Feb. 15th '03	w/e Sat. Nov. 9th '02	w/e Sat. Aug. 3rd '02	w/e Sat. April 27th '02
Extras 46	Extras 32	Extras 18	Extras 4
w/e Sat. Feb. 8th '03	w/e Sat. Nov. 2nd '02	w/e Sat. July 27th '02	w/e Sat. Apri 20th '02
Extras 45	Extras 31	Extras 17	Extras 3
w/e Sat. Feb. 1st '03	w/e Sat. Oct. 26th '02	w/e Sat. July 20th '02	w/e Sat. April 13th '02
Extras 44	Extras 30	Extras 16	Extras 2
w/e Sat. Jan. 25th '03	w/e Sat. Oct. 19th '02	w/e Sat. July 13th '02	w/e Sat. April 6th '02
Extras 43	Extras 29	Extras 15	Extras 1

Figure A.1 Sample page from the customer book

House No.	Street name *Church Lane*						
	Mon	Tues	Wed	Thur	Fri	Sat	Sun
2	*Mail*	*Mail*	*Mail*	*Mail*	*Mail*	*Mail*	*Sport*
4	*Ind*	*Ind*	*Ind*	*Ind*	*Ind*	*Ind Sat*	*Ind Sun*
6	*Guard*	*Guard*	*Guard*	*Guard*	*Guard*	*Times*	*Sun Times*
8	*Mail*	*Mail*	*Mail*	*Mail*	*Mail*	*Times*	*Sun Times*

Figure A.2 Sample page from the round book

	2002	**September**
	Sunday	**8**
restart	*14 Station Rd*	
restart	*103 Benny's Way*	
stop	*56 St John's Close*	
start	*12 St John's Close TES for 3 weeks*	

Figure A.3 Sample page from the diary

2. Data flow diagrams of Jays Newsagents

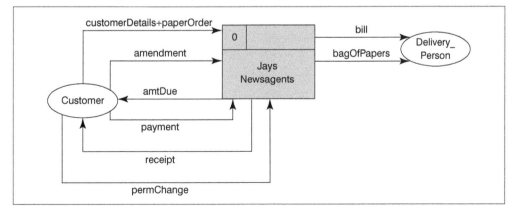

Figure A.4 Jays Newsagents: context diagram

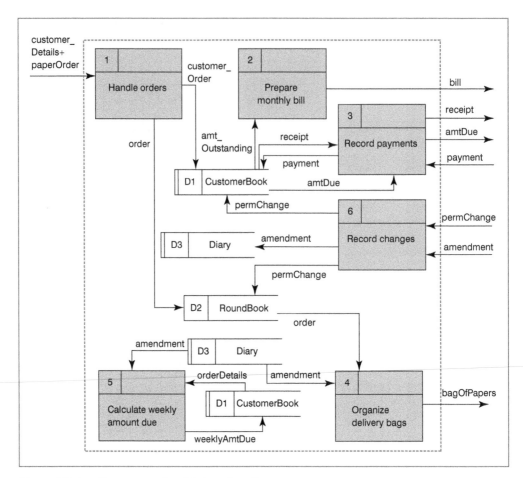

Figure A.5 Jays Newsagents: level 1 data flow diagram

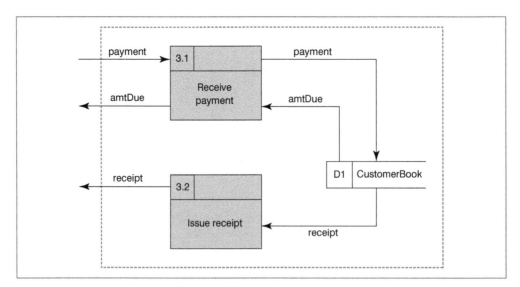

Figure A.6 Level 2 data flow diagram: expansion of process 3, Record payments

3. Partial data dictionary for Jays Newsagents

This data dictionary includes details of the data stores only.

```
CustomerBook        = *data store* {customerDetails +
                      paperRequirement +
                      deliveryCharge + totalAmountDue + {slip}}
customerDetails     = streetName + customerName + houseNo
paperRequirement    = {magazine + magPrice} + {localPaper +
                      localPrice} +
                      {periodical + periodicalPrice} +
                      {sundayPaper + sundayPrice} +
                      {dailyPaper + {day} + dailyPrice}
slip                = weekEndingDate + weeklyAmountDue +
                      (extraCharges) +
                      weekNumber
Diary               = *data  store* {date + {instruction + (comment)}}
instruction         = ["restart" | "stop" | "start"] + address
RoundBook           = *data store* {roadName + {houseNo +
                      {dailyOrder}}}
dailyOrder          = day + {paperName}
```

APPENDIX B: MIKE'S BIKES CASE STUDY

1. Case study description

Mike's Bikes is a bicycle hire company. The system it uses for keeping track of the bike hiring is as follows.

A customer goes to the reception desk and asks for the type of bike they want to hire, e.g. a woman's ten-speed racer in blue. The receptionist relays this information to a member of staff, who brings out a suitable bike from the storeroom.

The receptionist checks the bike number, which is painted on the side of the bike, and pulls out the index card for this bike from a card index box marked 'In Box'. The card (see Figure B.1) has on it details about the bike, such as its number, make, model, type (man's, woman's, child's) colour and frame size. The card also shows the daily hire rate and the deposit, and has space to record details about each hire transaction. The receptionist fills in details of the customer's name and address, and the start and return dates of the hire. She then works out the amount due, daily hire × number of days + deposit, and records this on the

Bike Number: 4591 Make: Scott Model: Atlantic Trail				
Type: woman's Colour: black Daily Rate: £8 Deposit: £50				
Frame size: 19				
Customer	**Start Date**	**Return Date**	**Paid**	**Extras**
Mrs V. Varty 16 St John's Road	31/08/02	2/09/02	£74	
Ms C. Wilson 112 Regent Street	9/09/02	12/09/02	£82	£8
Dr F. Green 67 Grange Road	4/10/02	4/10/02	£58	

Figure B.1 Mike's Bikes index card recording details of bike and hires

card. The amount due is paid by the customer, who then takes the bike and departs. The receptionist files the completed card in an index box marked 'Out Box'.

When the bike is returned, the customer tells the receptionist his or her name and the relevant card is extracted from the 'Out Box'. The details on the card are checked to see that they correspond with the bike that is being returned. The date of return is also checked; if the return is late, the customer is required to pay an extra charge.

The bike is then examined by a mechanic to check for any damage. If the bike is in good condition the receptionist returns the deposit to the customer.

2. Data flow diagrams of Mike's Bikes

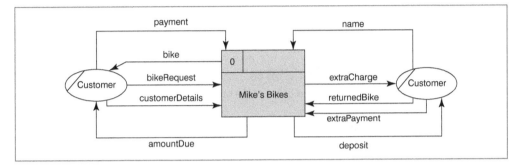

Figure B.2 Mike's Bikes: context diagram

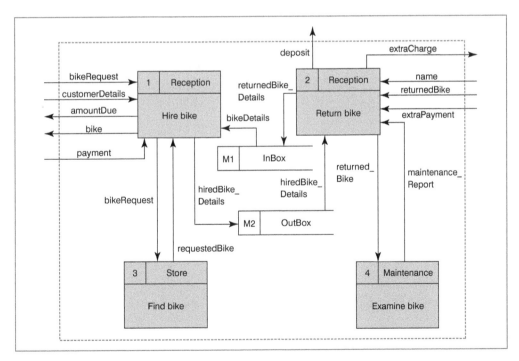

Figure B.3 Mike's Bikes: level 1 current physical data flow diagram

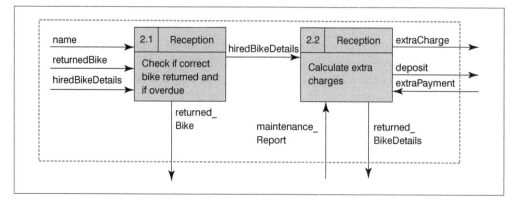

Figure B.4 Mike's Bikes: level 2 current physical data flow diagram

3. Partial data dictionary for Mike's Bikes

This data dictionary includes details of the data stores only.

```
         InBox = *data store* {indexCard}
        OutBox = *data store* {indexCard}
   bikeDetails = bike# + make + model + type + colour + frameSize
       charges = dailyRate + deposit
customerDetails = name + address
hireTransactions = customerDetails + startDate + returnDate +
                  amountPaid + (extras)
     indexCard = bikeDetails + charges + {hireTransactions}
          type = ["man's"|"woman's"|"child's"]
```

ANSWERS TO SELECTED EXERCISES

Chapter 1

Exercise 1.1

Shower
Purpose: to provide a spray of water at a chosen temperature.
Elements: water, electric heater, method of controlling temperature and flow, cubicle.
Boundary and environment: the cubicle provides the boundary; everything outside it is the environment.

Car
Purpose: to provide a means of conveying people from one place to another.
Elements: engine, body, seats, steering system, electrical system, etc.
Boundary and environment: provided by the body of the car; everything outside it is the environment.

School library
Purpose: to lend books and other publications to school pupils and staff.
Elements: books, other publications, room, shelves, shelving system, index, method of recording loans and returns, etc.
Boundary and environment: inside the system are the library room and all its elements; a school library system would be unlikely to include connections to other libraries. Pupils and staff are regarded as users of the library, rather than part of the system.

Exercise 1.2
Refer to the bibliography for suitable books. Pressman's *Software Engineering: A Practitioner's Approach* (2004) covers different approaches to the software life cycle in Chapter 2 (see Bibliography for full details). Some of the most widely used life cycle models are the Waterfall, the Spiral and prototyping.

Exercise 1.3
Discussion on this question should include factors such as the amount of money available, the time allowed for development, the wishes of the client, the preferences of potential users, the nature of the local area. Decisions about the system boundary should consider both the physical area to be covered by the system and the topics to be included, such as maps, transport, local walks and cycle paths, medical centres, shopping and entertainment.

Chapter 2

Exercise 2.1

Points to be discussed could include the following.

- The software crisis proved that new methods were needed.
- Large systems will typically involve discussions with several users and will be developed by a team.
- Need for a systematic approach to capturing user requirements.
- Need to co-ordinate teamwork, common approach, common vocabulary.
- Importance of milestones for project management – estimating and controlling a project.

Exercise 2.2

Documents in the Just a Line system:

- list of card designs
- price list
- customer order form and delivery note
- supplier order form
- list of stock in hand
- supplier invoice
- cash book.

Exercise 2.3

Points to include are as follows.

- A method provides a mechanism to help the developer divide the problem area into manageable portions.
- A method provides compatible techniques to build useful models and the ability to verify and cross-check each step of the development.
- Distinct phases, sub-phases and tasks.
- Good project management guidelines, control and evaluation.
- Many different types of system exist; no one method will be suitable for all.
- Rigid adherence to a method may result in unnecessary work being done, and in useful techniques being ignored because they are not part of the method used.

Exercise 2.4

Stage	Deliverable	
1	Problem definition	Problem definition – statement of problems, scope and objectives of new system
2	Feasibility study	Feasibility study report – analysis of viability of project, rough cost–benefit analysis
3	Requirements engineering	Specification of requirements – logical model of required system may include data model, required logical data flow diagrams, data dictionary, process definitions
4	Design	Technical design specification, includes program specifications, hardware specifications, cost estimates and an implementation schedule
5	Implementation	Working system, includes program listings and documentation, test plan, hardware, operating procedures, clerical procedures
6	Installation (if included as a separate stage)	Operational system, trained users
7	Maintenance	Operational system, modified and documented as required

Exercise 2.5

Changes to the problem definition should include the problem of wholesale orders, and the objective of handling very large orders from companies that will demand a fast, efficient and reliable service. The feasibility study should investigate how Just a Line is going to cope with this new type of customer.

Chapter 3

Exercise 3.1

■ Information that is already structured in lists or forms, e.g. the supplier's catalogue.

■ Information about company procedures and how certain tasks are carried out at present, e.g. how orders are handled and how the various bits of paper are filed.

■ Measurements such as the number of customers or the average size of an order, e.g. free supplier delivery if Just a Line order is worth more than £500.

■ Problems that the client has identified in the current system, e.g. hit-and-miss stock control.

■ Definite requirements for the new system, e.g. the mailing list.

■ Information that is not stated directly, but where there are definite 'vibes'. Examples of this might be where the clients complain that they are always rushed when the supplier's order comes in, whereas what is actually happening is that the supplier always delivers late; 'Customers are so chatty' meaning 'It takes ages to find out what exactly they want'; 'It's just an overall feeling of being disorganized' meaning 'We're in a right mess'.

Exercise 3.2

There are many more questions that could have been asked. Here are some examples.

■ How long do you keep past orders?

■ Approximately how many customers do you have at the moment?

■ What is your target number of customers?

■ How much money do you have to spend on the system?

■ How many orders do you handle on average each week?

■ To what extent do you want orders from large stores to become the main part of your business?

■ What happens when you have a holiday?

Exercise 3.3

Town Council Car Park Survey.

In order to improve the organization of this car park, we are carrying out a survey to see how this can best be achieved. If you are a regular user of the car park, please spare a few minutes to answer the questions below and place the form in the box. This is your car park and we would like your ideas on how it should be organized.

Please circle your answers to the following questions.

1. How many times a week on average do you use this car park?
   never/once or twice/two or three times/every day/more than once a day

2. Which car park entrance do you normally use to come into the car park?
   main entrance/back entrance/either

3. Which car park exit do you normally use to leave the car park?
   main exit/back exit/either

4. For which of the following purposes do you use the car park?
   shopping/work/cinema/other

5. Do you have problems finding a space in the car park?
   never/sometimes/frequently/always

6. If you have problems finding a space, when does this usually happen?
   at the start of work/during the morning/lunchtime/during the afternoon/
   at the end of the day/no regular pattern

7. Would you like to see more security in the car park?
   yes/no/don't mind

8. Do you think that the lighting in the car park is adequate?
   yes/no/don't know

9. How much would you be prepared to pay per hour to park here if the facilities were improved (e.g. more spaces, better lighting)?
   20p/50p/£1.00/not prepared to pay anything

10. Please note below any suggestions that you have for improving the organization of the car park.

   _____

   _____

   If you would like to receive information about the results of this survey, please fill in details of your name and address.

   Your name:

   _____

   Your address:

   _____

   _____

   _____

   Thank you for completing this questionnaire.

Chapter 4

Exercise 4.1

The balancing rule is not infringed. Notice that flows to and from the data stores Dl and D2 are shown inside the boundary on the level 2 diagram.

Exercise 4.2

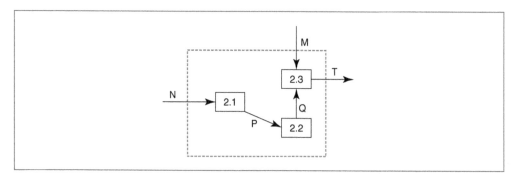

Figure E4.2 Correct version of child diagram (from Figure 4.8)

Exercise 4.3(a)

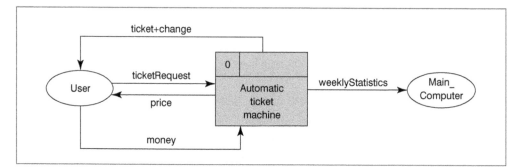

Figure E4.3(a) Automatic ticket machine: context diagram

Exercise 4.3(b)

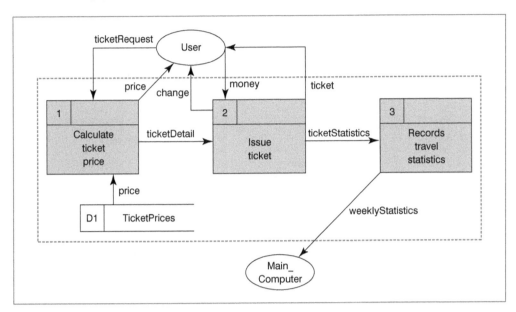

Figure E4.3(b) Automatic ticket machine: level 1 current logical data flow diagram

Exercise 4.4

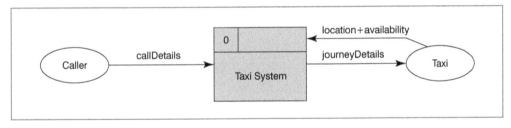

Figure E4.4(a) Taxi system: context diagram

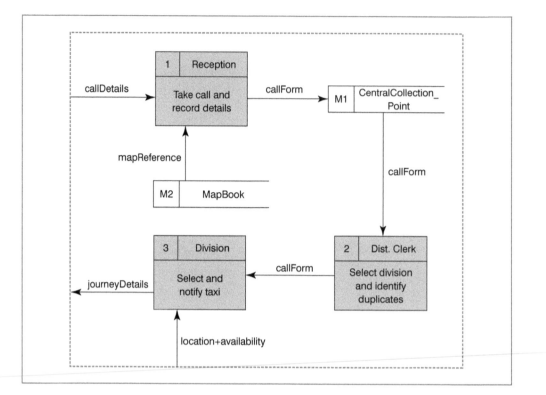

Figure E4.4(b) Taxi system: level 1 current physical data flow diagram

Exercise 4.5

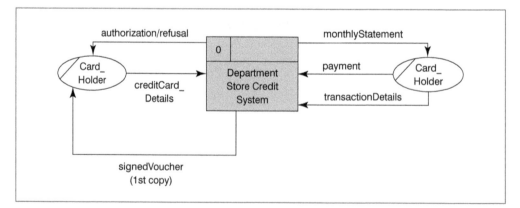

Figure E4.5(a) Credit system: context diagram

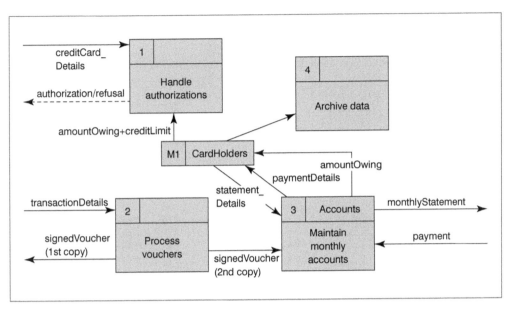

Figure E4.5(b) Credit system: level 1 current physical data flow diagram

Exercise 4.6

Note that in Figure E4.6(a) Agent is shown as an external entity – this is a contrivance to get **valuationNotes** into the system. Another approach would be to have the **valuationNotes** coming directly from the external entity **Seller**, but this seems unsatisfactory.

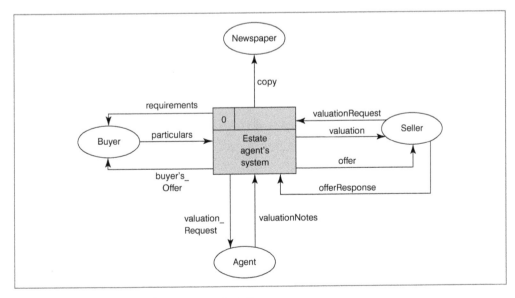

Figure E4.6(a) Estate agent's system: context diagram

Note that, in Figure E4.6(b):

- the data flow **particulars** appears twice on the diagram
- a two-way arrow is used to indicate that data is both retrieved and updated.

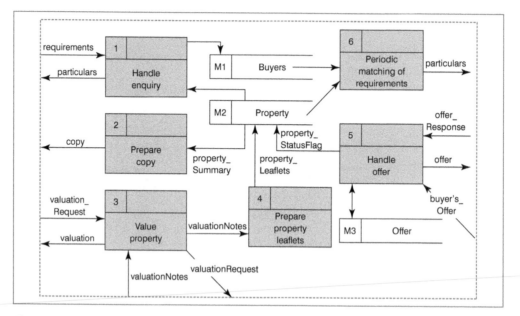

Figure E4.6(b) Estate agent's system: level 1 current physical data flow diagram

Exercise 4.7

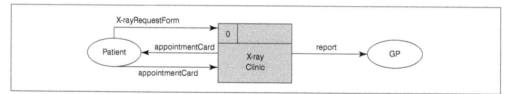

Figure E4.7(a) X-ray clinic: context diagram

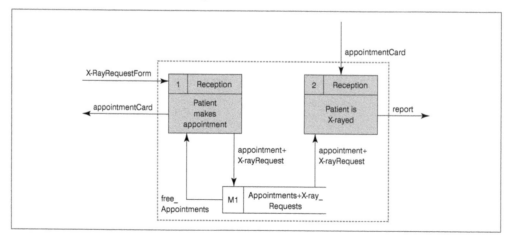

Figure E4.7(b) X-ray clinic: level 1 current physical data flow diagram

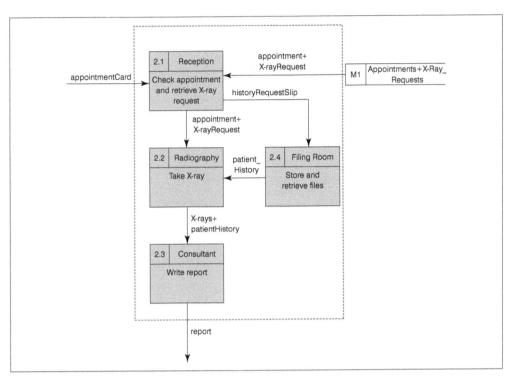

Figure E4.7(c) X-ray clinic: level 2 current physical data flow diagram

Note that the data store M1 **Appointments + X-rayRequests** is shown outside the boundary. This indicates that the data store has already appeared in a higher-level diagram.

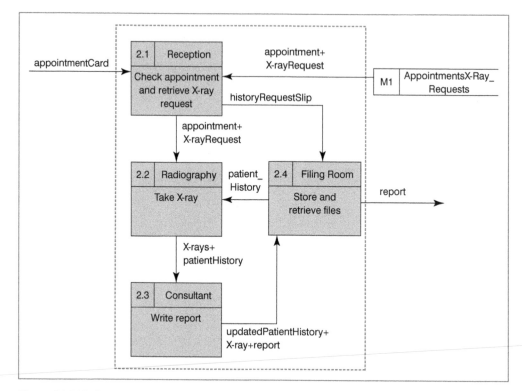

Figure E4.7(d) X-ray clinic: level 2 current physical data flow diagram (revised version)

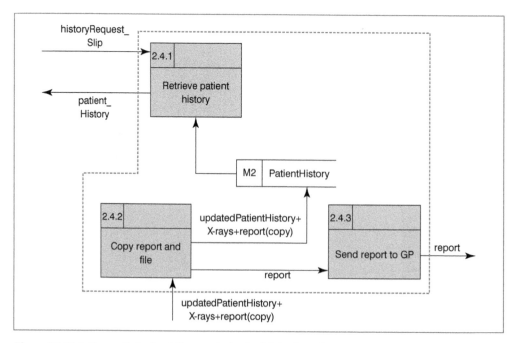

Figure E4.7(e) X-ray clinic: level 3 current physical data flow diagram, expansion of process 2.4

Exercise 4.8

1. The data flow between process 1.3 and the external entity H.O. should be labelled.
2. The flow between processes 1.2 and 1.3 has no label – it seems to be modelling flow of control, which is not allowed.
3. One of the flows from process 1.2 is labelled **sendOutBill**. This contains an active verb, indicating that the flow is being used to model an activity.
4. The diagram has no system boundary.
5. The data flow between data stores DI and D2 is not allowed – it must go via a process.
6. The process **Handle stock requirements** has no reference number.
7. The duplication indicator should appear on both occurrences of the data store **Roundsman'sBook**.
8. The flow **newPrices** from the external entity H.O. to the data store **Roundsman'sBook** should go via a process.
9. The data stores should have a reference number prefix M not D.

Exercise 4.9

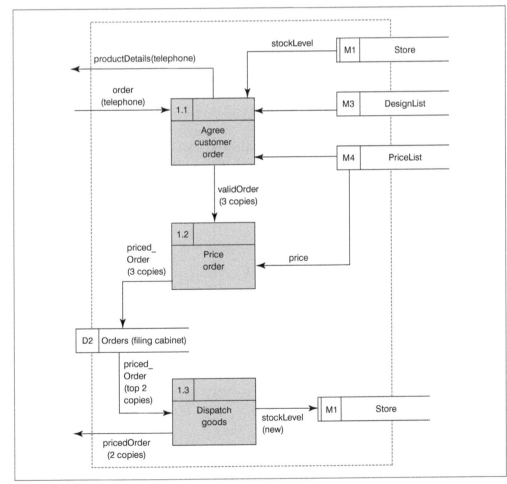

Figure E4.9 Just a Line CPDFD: level 2 expansion of process 1

Exercise 4.10

Questions might include the following.

- When and how is the design list updated? When and how is the price list updated?
- Are the petty cash and banking transactions to be excluded from the project?
- Exactly what information is recorded in the cash book?
- Exactly what information is recorded in the design list and price list? Are customers charged for delivery or postage?
- Do the designs have code numbers or are they identified by description only?

Exercise 4.11

Details of the products (taken from the data store **ProductInformation**) are supplied to the customer; orders are sent in from the customer. The orders are priced (from information in the data store **ProductInformation**) and stored in the data store Orders until they are ready to be made up. When the order is made up it is delivered to the customer with a copy of the order serving as **invoice** + **deliveryNote**. At this point stock levels are updated in the data store **ProductInformation**.

The accounting procedure will be provided with the **customerPayment** when it arrives and with a copy of the **pricedOrder**.

Ordering of stock from the supplier is done using information about stock levels, codes and product descriptions in the data store **ProductInformation**, and information about the quantities required by outstanding customer orders. This information is compiled and written on a **stockOrder** form, which is sent to the supplier. The supplier sends Sue and Harry the stock they ordered with a **supplierInvoice**. The **supplierInvoice** is used to check the goods supplied and update the stock levels in the data store **ProductInformation**. A signed copy is returned as **confirmationOfDelivery**. A copy of the **supplierInvoice** is supplied to the accounting procedure. This in turn generates a **supplierPayment**.

Exercise 4.12

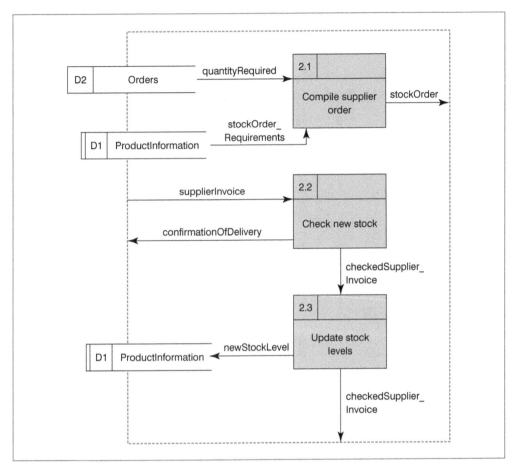

Figure E4.12 Just a Line CLDFD: level 2 expansion of process 2

Exercise 4.13

Points include the following.

- CPDFDs are useful if the system seems very complicated and users give conflicting accounts.
- CPDFDs see the system in the user's terms (forms in triplicate, pink cards, filing cabinets, telephone calls, etc.). Less sophisticated users may find this easier to follow than the more abstract CLDFD.
- Drawing a CPDFD may influence the system developer's design of the new system; it may prove hard to break away from the old design.
- Drawing both a CPDFD and a CLDFD is time-consuming.
- Experienced developers may draw data flow diagrams while the user is describing the system. This helps the user to understand the notation, and helps the developer to realize where there are gaps in his or her understanding of the system and to ask appropriate questions.

Exercise 4.14(b)

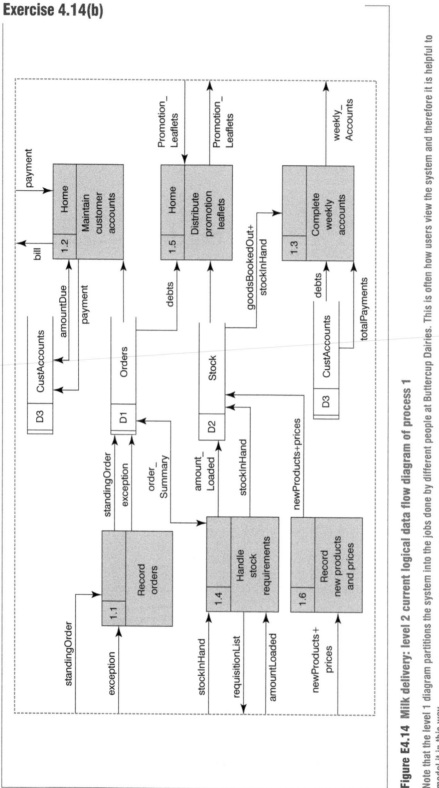

Figure E4.14 Milk delivery: level 2 current logical data flow diagram of process 1

Note that the level 1 diagram partitions the system into the jobs done by different people at Buttercup Dairies. This is often how users view the system and therefore it is helpful to model it in this way.

On the level 2 CPDFD the jobs are clearly seen as those done at home, those done on the round and those done at the depot; i.e. jobs are grouped according to where they are done. Partitioning the system according to location is another typical feature of modelling done at the current physical stage. Both types of partitioning are abandoned at the current logical stage as it becomes irrelevant who does a particular job, or where it is done.

Exercise 4.16

Red	Y	Y	N	N	
Amber	N	Y	N	Y	
Green	N	N	Y	N	
Go			X		
Stop	X				
Prepare to stop				X	
Prepare to go		X			

Figure E4.16 Decision table

Exercise 4.17(a)

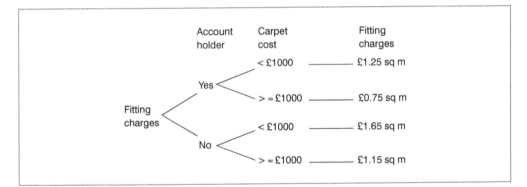

Figure E4.17(a) Decision tree

Exercise 4.17(b)

```
if customer is accountHolder then
    if carpetCost < £1000 then
            fittingCharge = £1.25 per sq. m.
    else (*carpetCost >= £1000*)
            fittingCharge = £0.75 per sq. m.
    else (*customer not accountHolder*)
            if carpetCost < £1000 then
                    fittingCharge = £1.65 per sq. m.
    else (*carpetCost >= £1000*)
            fittingCharge = £1.15 per sq. m.
```

Exercise 4.18

(a)

Rules	1	2	3	4	5	6	7	8
raining	Y	Y	Y	Y	N	N	N	N
November	Y	Y	N	N	Y	Y	N	N
windy	Y	N	Y	N	Y	N	Y	N
stay at home	X	X						
take umbrella			X	X				
wear coat			X		X		X	
wear hat				X		X		X
put on central heating	X				X			

Notes

1. The action 'stay at home' is considered to be the opposite of 'going to work' so it does not require a separate entry in the set of actions.
2. If two vertical rules have the same outcome and conditions that vary in only one respect, they can be combined with the varying conditions declared to be immaterial. In this case rules number 6 and 8 can be combined: if it is not raining and not windy, Miss Take wears her hat whether or not it is November. See version (b).

(b)

Rules	1	2	3	4	5	6	7	
raining	Y	Y	Y	Y	N	N	N	
November	Y	Y	N	N	Y	–	N	
windy	Y	N	Y	N	Y	N	Y	
stay at home	X	X						
take umbrella			X	X				
wear coat			X		X		X	
wear hat				X		X		
put on central heating	X				X			

Exercise 4.19(a)

```
Process 1: Calculate ticket price
get ticketRequest from User
get price of destination from Dl TicketPrices
display price to User
send ticketDetail to process 2: Issue ticket
```

Exercise 4.19(b)

```
Process 2: Issue ticket
get ticketDetail from process 1: Calculate ticket price
get money from User (*until money >= price*)
if money > price then
        give change to User
        issue ticket to User
        send travelStatistics to process 3: Record travel
        statistics
```

Chapter 5

Exercise 5.3

```
change = *amount of money returned if money > price*
money = *amount input by User*
price = *price of ticket*
ticket = date + issuingStation + destination + price + ticketType
ticketDetail = destination + ticketType + price
TicketPrices = *data store* destination + {ticketType + price}
ticketRequest = destination + ticketType
ticketStatistics = ticketRequest
weeklyStatistics = date + {destination + {ticketType + no.Sold}}
```

Exercise 5.4

```
statement = stmtNo + cardNo + stmtDate + custName + custAddr +
            {transactionDate + code + (debitAmt) + (creditAmt)} +
            paymentDue + (creditBalance)
code = [department | reference]
```

Alternatively, the document can be described more elegantly using data structures:

```
statement = stmtHeader + {stmtLines} + paymentDue + (creditBalance)
stmtHeader = stmtNo + cardNo + stmtDate + custName + custAddr
stmtLines = transactionDate + code + (debitAmt) + (creditAmt)
code = [department | reference]
```

Exercise 5.5

```
standingOrder = customerDetails + standingOrderDetails
customerDetails = customerAddress + customerName + deliveryPoint
standingOrderDetails = {day + {product + quantity}}
exception = customerAddress *used as identifier* + {date + {product
            + quantity *may be positive or negative*}}
Roundsman'sBook = *data store* roundNumber + {customerPage}
customerPage = customerDetails + standingOrderDetails +
               {weeklyAccountDetails)
weeklyAccountDetails = {week'sException} + amountDue + (payment)
week'sException = date + {product + quantity)
```

Chapter 6

Exercise 6.1

(This solution uses # to mean number. You may have preferred to use No.)

(a) Candidate keys are emp# (or empNo), name and startDate; scale, dept# and office# are not candidate keys because none of them is unique to each employee; ext# cannot be a candidate key because I. Oxford does not have an extension number – this attribute, therefore, can have null values.

(b) emp# will make the best primary key because it uniquely identifies each employee, its allocation is under the control of the system users, it never has a null value and it is not liable to change; name would not make a good primary key as an employee's name can change and it is quite possible, over a period of time, that the company might employ two people with the same name, although that is not the case at the moment. startDate would not make a good primary key because it is quite possible for two employees to start work on the same day.

Exercise 6.2

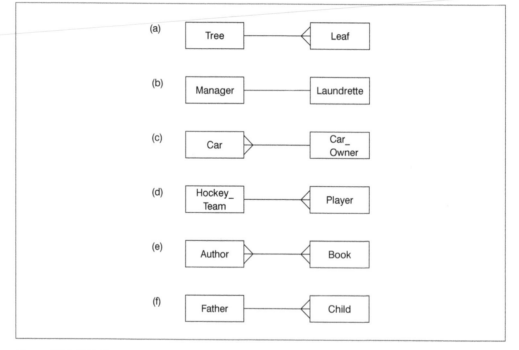

Figure E6.2 E-R diagrams

Exercise 6.3

(a) True

(b) False

(c) False

(d) True

Exercise 6.4(a)

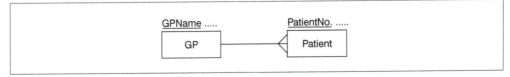

Figure E6.4(a) A GP has many patients, a patient has one GP

Exercise 6.4(b)

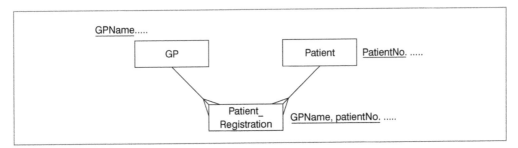

Figure E6.4(b) A GP has many patients, a patient may be registered with many GPs

Exercise 6.5

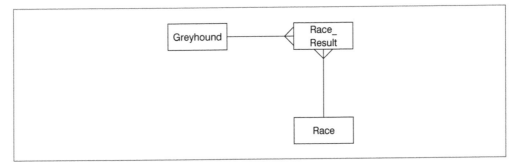

Figure E6.5 Introduction of intersection entity

Exercise 6.6 (a) and (b)

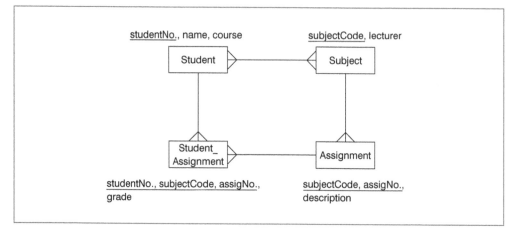

Figure E6.6 E-R model showing relationships of Students, Assignments and Subjects

Exercise 6.7
The grade depends on the mark.

Exercise 6.8
The lecturer depends on the student.

Exercise 6.9
```
Customer(customer#, name, address)
VideoRental (customer#, videoRef., dateBorrowed, dateDueBack)
```

Exercise 6.10
```
Employee (employee#, name, department)
Languages (employee#, language, level)
```

Exercise 6.11
```
Pupil (pupilName, class, height, weight)
PupilTime (pupilName, distance, time)
```

Exercise 6.12
```
1NF    Student (student#, studentName)
       StudentOption (student#, optionCode, optionName, lecturer,
       studentGrade)
2NF    Student (student#, studentName)
       Option (optionCode, optionName, lecturer)
       StudentOption (student#, optionCode, studentGrade)
```

Exercise 6.13

```
1NF    Borrower (borrower#, borrowerName, borrowerAddress)
       BookLoan (borrower#, book#, title, author, ISBN, dateBorrowed,
       dateDueBack)
2NF    Borrower (borrower#, borrowerName, borrowerAddress)
       Book (book#, title, author, ISBN)
       BookLoan (borrower#, book#, dateBorrowed, dateDueBack)
```

This assumes that the BookLoan record is deleted when the book is returned – see discussion on the library system in Chapter 6

Exercise 6.14

```
1NF    Room (room#, roomType, roomPrice)
       Booking (room#, guestName, guestAddress, dateOfArrival,
       dateOfDeparture)
2NF    Room (room#, roomType, roomPrice)
       Guest (guestName, guestAddress)
       Booking (room#, guestName,
       dateOfArrival, dateOfDeparture)
```

Exercise 6.15

```
1NF    Pupil (pupilName, dateOfBirth)
       PupilReader (pupilName, reader#, readerTitle, vocabularyLevel)
2NF    Pupil (pupilName, dateOfBirth)
       Reader (reader#, readerTitle, vocabularyLevel)
       PupilReader (pupilName, reader#)
```

Exercise 6.16

```
1NF    Doctor (doctorCode, doctorName)
       Referral (doctorCode, patientName, patientAddress,
       patientPhone#, diagnosisName, dateOfReferral,
       consultant, hospital)
2NF    Doctor (doctorCode, doctorName)
       Patient (patientName, patientAddress, patientPhone#)
       Referral (doctorCode, patientName, diagnosisName,
       dateOfReferral, consultant, hospital)
3NF    Doctor (doctorCode, doctorName)
       Patient (patientName, patientAddress, patientPhone#)
       Referral (doctorCode*, patientName*, diagnosisName,
       dateOfReferral, consultant*)
       Consultant, (consultant, hospital)
```

doctorCode, patientName and consultant, are foreign keys in Referral.

Exercise 6.17

```
1NF    MultipleChoiceTest (testID, topic, level,
       estimatedTimeToComplete, authorName, department, email)
       MultipleChoiceQuestion (testID, question#, questionType)
2NF    as 1NF
3NF    MultipleChoiceTest (testID, topic, level,
       estimatedTimeToComplete, authorName*)
       MultipleChoiceQuestion (testID*, question#, questionType)
       Author (authorName, department, email)
```

authorName is a foreign key in MultipleChoiceTest; testID is a foreign key in MultipleChoiceQuestion.

Exercise 6.18

```
1NF    Consultant (consultant#, consultantName, consultantPhone#)
       ConsultantDay (consultant#, day, unit)
       ConsultantList (consultant#, day, time, patient#, patientName,
       patientAddr, GP#, GPName, GPAddr, GPPhone#)
2NF    All of the 1NF entities are also in 2NF
3NF    Consultant (consultant#, consultantName, consultantPhone#)
       ConsultantDay (consultant#*, day, unit)
       ConsultantList (consultant#*, day*, time, patient#*)
       Patient (patient#, patientName, patientAddr, GP#*)
       GP (GP#, GPName, GPAddr, GPPhone#)
```

Consultant# is a foreign key in ConsultantDay; consultant#, day and patient# are foreign keys in ConsultantList; GP# is a foreign key in Patient.

Exercise 6.19

```
UNF    Sales (salesperson#, name, area, (customer#, customerName,
       warehouse#, warehouseSite, totalSales)
1NF    Salesperson (salesperson#, name, area)
       Salesperson'sCustomer (salesperson#, customer#, customerName,
       warehouse#, warehouseSite, totalSales)
2NF    Salesperson (salesperson#, name, area)
       Salesperson'sCustomer (salesperson#, customer#)
       Customer (customer#, customerName, warehouse#, warehouseSite,
       totalSales)
3NF    Salesperson (salesperson#, name, area)
       Salesperson'sCustomer (salesperson#*, customer#*)
       Customer (customer#, customerName, warehouse#*, totalSales)
       Warehouse (warehouse#, warehouseSite)
```

salesperson# and customer# are foreign keys in Salesperson'sCustomer; warehouse# is a foreign key in Customer.

Exercise 6.20

```
UNF    TreatmentCard (animal#, name, breed, block#, type, blockName,
       (startDate, drugCode, drugName, dosage, condition, length))
1NF    Animal (animal#, name, breed, block#, type, blockName)
       Treatment (animal#, startDate, drugCode, drugName, dosage,
       condition, length)
2NF    Animal (animal#, name, breed, block#, type, blockName)
       Treatment (animal#, startDate, drugCode, dosage, condition,
       length)
       Drug (drugCode, drugName)
3NF    Animal (animal#, name, breed, block#*)
       Block (block#, type, blockName)
       Treatment (animal#*, startDate, drugCode*, dosage, condition,
       length)
       Drug (drugCode, drugName)
```

block#* is a foreign key in Animal; animal# and drugCode are foreign keys in Treatment

Exercise 6.21

```
UNF    MemberOrders (member#, memberName, memberAddress,
       (dateOrdered, (title#, title, publisherCode, publisherName)))
1NF    Member (member#, memberName, memberAddress)
       Member'sOrder (member#, dateOrdered)
       Member'sOrderTitle (member#, dateOrdered, title#, title,
       publisherCode, publisherName)
2NF    Member (member#, memberName, memberAddress)
       Member'sOrder (member#, dateOrdered)
       Member'sOrderTitle (member#, dateOrdered, title#)
       BookTitle (title#, title, publisherCode, publisherName)
3NF    Member (member#, memberName, memberAddress)
       Member'sOrder (member#*, dateOrdered)
       Member'sOrderTitle (member#*, dateOrdered*, title#*)
       BookTitle (title#, title, publisherCode*)
       Publisher (publisherCode, publisherName)
```

In Member'sOrderTitle, member# is a foreign key referring to Member; member# with dateOrdered is a composite foreign key referring to Member'sOrder; title# is a foreign key referring to BookTitle.

Chapter 7

Exercise 7.1

(a) Customer places order. **(b)** No. **(c)** Any number of times, including zero. **(d)** No. **(e)** The order must either have been delivered or cancelled.

Exercise 7.2

Figure 7.17(b) is correct.

Exercise 7.3

It is not correct to draw a repeat node and an event node at the same level from the same box.

Exercise 7.4

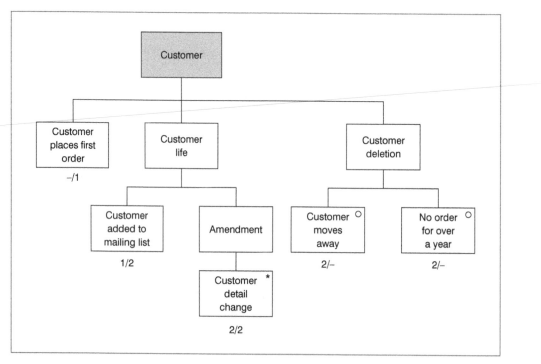

Figure E7.4 Customer entity life history for the Just a Line required system

Exercise 7.5

When an order is filled an occurrence of the Invoice entity is created. During the life of the invoice one of two things will happen, either the invoice is paid or the goods are returned. Finally, the details of the invoice are deleted.

Exercise 7.6

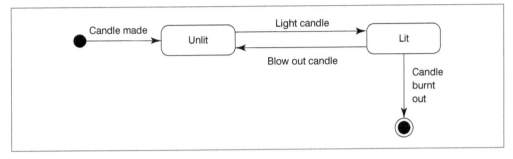

Figure E7.6 State diagram for a candle

Exercise 7.7

(a) The tank moves into the Full state.

(b) The tank stays in the Part full state.

(c) The warning light is turned on when the tank is part full and then all the petrol is used up.

(d) Three events can happen:

- petrol may be added, but not to capacity
- petrol may be added to fill the tank
- the car may be written off.

Exercise 7.8

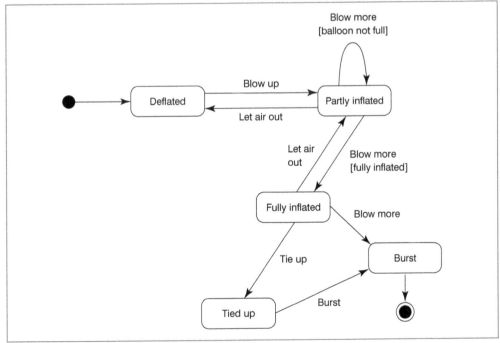

Figure E7.8 Modified state diagram for the Balloon entity

Exercise 7.9

To begin with, the machine is idle. When the user selects the type of crisps he or she wants, the machine requests payment. The user inserts coins until the correct amount, or more, has been entered (the machine does not give change). The packet of crisps is then dispensed. The transaction is now complete and the machine returns to the idle state.

Exercise 7.10

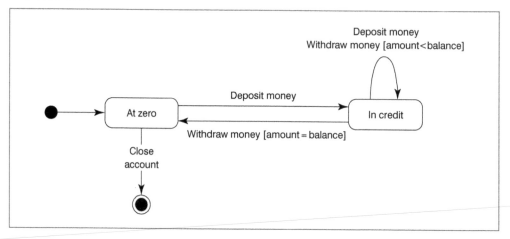

Figure E7.10 State diagram for a child's bank account

Exercise 7.11

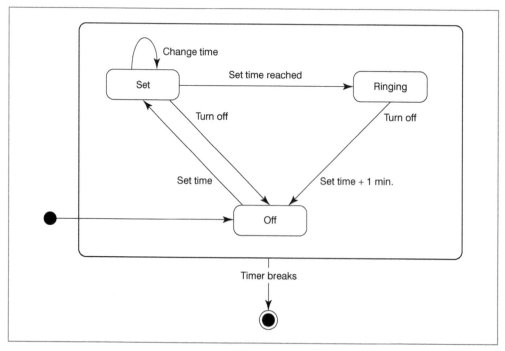

Figure E7.11 State diagram for a kitchen timer

Chapter 8

Exercise 8.1

Supplier

Supplier No.	Supplier Name	Supplier Address	Supplier Phone
S01	Bentons	Unit 4, Industrial Estate, Hansford SC7 5LH	01483 822976
S02	J.J. Lawton	14–16 High Road, Boxeth SC8 1DR	01485 324773

Exercise 8.2

A one-to-many relationship is implemented by including the primary key from the table of the entity at the '!' end of the relationship in the table of the 'many' entity. In this example, the Order table would include the primary key of Customer.

Exercise 8.3

In the case of a many-to-many relationship, it is necessary to construct a separate table to implement the relationship in the database. In this example, the many-to-many relationship Customer to Delivery is split into two one-to-many relationships, Customer to CustomerDelivery and Delivery to CustomerDelivery, and three tables are produced: Customer, Delivery and CustomerDelivery. The primary key of the table CustomerDelivery is made up of the primary keys of the other two tables.

Answers to SQL exercises

Exercise 8.7

```
INSERT INTO DVDLoan
VALUES ('C191', 'D406', '05/05/2005', 2);
```

Exercise 8.8

```
UPDATE DVD
SET DailyHireRate = DailyHireRate * 1.025;
```

Exercise 8.9

```
DELETE FROM DVDCustomer
WHERE CustomerFirstName = 'Brian'
AND CustomerName = 'Wells';
```

Exercise 8.10

```
SELECT *
FROM DVD;
```

Exercise 8.11

```
SELECT DVDNo, DVDName
FROM DVD
WHERE Category = 'Historical';
```

Exercise 8.12

```
SELECT DISTINCT Category
FROM DVD;
```

Exercise 8.13

```
SELECT COUNT (*) AS Total
FROM DVDCustomer;
```

Exercise 8.14

```
SELECT COUNT (*) AS Juniors
FROM DVDCustomer
WHERE Status = 'Junior';
```

Exercise 8.15

```
SELECT CustNo, DVDNo, DateIssued, DaysPaid + 1 AS ExtraDay
FROM DVDLoan;
```

Exercise 8.16

```
SELECT MIN (DailyLoanRate) AS MinRate
FROM DVD;
```

Exercise 8.17

```
SELECT AVG (DailyLoanRate) AS Average
FROM DVD;
```

Exercise 8.18

```
SELECT *
FROM DVD
ORDER BY DailyLoanRate ASC;
```

Exercise 8.19

```
SELECT CustFirstName, CustName, CustPhone
FROM DVDCustomer
ORDER BY CustName ASC;
```

Exercise 8.20

```
SELECT DVDName
FROM DVD, DVDLoan
WHERE DVD.DVDNo = DVDLoan.DVDNo;
```

Exercise 8.21

```
SELECT DVDCustomer.CustNo, CustName
FROM DVDCustomer, DVDLoan
WHERE DVDCustomer.CustNo = DVDLoan.CustNo;
```

Exercise 8.22

```
SELECT DISTINCT DVDCustomer.CustNo, CustName
FROM DVDCustomer, DVDLoan
WHERE DVDCustomer.CustNo = DVDLoan.CustNo;
```

Exercise 8.23

```
SELECT DVDCustomer.CustNo, CustName
FROM DVDCustomer, DVD, DVDLoan
WHERE DVDCustomer.CustNo = DVDLoan.CustNo
AND DVDLoan.DVDNo = DVD.DVDNo
AND Status = 'Junior'
AND Category = 'Animation';
```

Exercise 8.24

```
SELECT DVDName
FROM DVD, DVDLoan, DVDCustomer
WHERE DVDCustomer.CustNo = DVDLoan.CustNo
AND DVDLoan.DVDNo = DVD.DVDNo
AND Status = 'Senior citizen';
```

Exercise 8.25

Display in alphabetical order the customer numbers and names of customers who have paid for less than two days.

CustNo	CustName
C517	Davis
C487	Jones
C522	Watson

Exercise 8.26

Display the names of the DVDs in the 'Thriller' category that have been borrowed by customers who are not juniors.

DVDname
The Pledge
Psycho
The Others

Chapter 9

Exercise 9.1

Examples of valid data:

- 46 72 9 501. (random whole numbers)
- 34 34 34 8 33. (more than one instance of largest number)
- 34 5 34 26 13. (largest number appears more than once, but not consecutively)
- 5. (only one number input)

Examples of invalid data:

- 4 8.7 32 1. (real number included in series)
- 4 8 2F 22. (hexadecimal number included in series)
- 4 8 M 22. (letter included in series)
- 4 & 8 22. (non-alphanumeric character included)
- 4 8 54 22 (no full stop to terminate series)
- 0 123456. (0 is not a positive number)

Exercise 9.2

School library system

- Probably not large enough for phased implementation or pilot running, although could initially be introduced for one class only.
- Unlikely that there are enough staff to cope with the extra load of parallel running.
- If the system breaks down it is inconvenient, but not disastrous, so direct changeover is probably the most appropriate method.

GPs' system

- Pilot running a possibility, using selected patients only.
- There must be a back-up for the system, so direct changeover is too risky.
- Parallel running is probably the best method, as long as the surgery staff can cope with the increased workload.

Dairy system

- Probably not worth the extra effort of parallel running.
- Phased implementation possible.
- Direct implementation a possibility as long as the client understands the risks involved.
- Pilot running, using one round only, probably the most suitable option.

Chapter 10

Exercise 10.1

The following points should be included.

- A team of developers will be needed (potential problems of group work).
- Must have a project leader/manager.
- Organization and communication among team members take up time.
- Probably more than one user.
- More partitioning of the problem required.
- Many tasks to be scheduled.
- Task dependencies and the critical path become more important.
- More resources to be organized and allocated.

Exercise 10.2

Some project management will be needed, but too much could be overkill on a small project. Points raised could include:

- team size (small)
- whether a separate project manager is necessary
- tasks need to be defined and scheduled
- resources need to be organized
- milestones should be set
- the progress of the project needs to be monitored and controlled.

Exercise 10.4

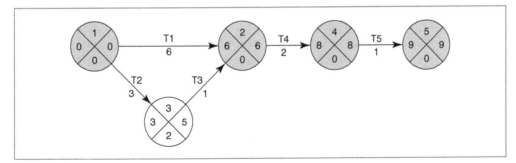

Figure E10.4 Milestones on the critical path: 1, 2, 4, 5

Exercise 10.5

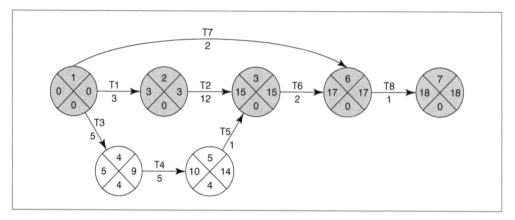

Figure E10.5 Milestones on the critical path: 1, 2, 3, 6, 7

Exercise 10.7

These are some of the issues to consider in these situations. Remember that there is generally no single right answer.

a) Is this an invasion of privacy under the Computer Misuse Act, or do the computer and files belong to the company?

The manager requested you to access the computer, so is it his responsibility?

What will be the effect on the company, its employees and the client if the project continues to run late?

You should tell the sick employee that the files have been copied, but should you tell him about what you have discovered?

Should you tell your manager about the files from the extreme right-wing organization?

b) The BCS Code of Conduct, Section 1, states that: 'You shall carry out work or study with due care and diligence in accordance with the relevant authority's requirements, and the interests of system users. If your professional judgement is overruled, you shall indicate the likely risks and consequences.'

Do you have the expertise and experience to recognize a genuine problem?

If you cannot convince your manager, should you go higher up in the company?

Should you talk to the client?

What sort of records should you be keeping of the situation?

What are the consequences of releasing a potentially faulty system?

c) This is a similar situation to the one in (b), but here the system is safety-critical and, if faulty, could be life threatening.

Should you talk directly to the hospital as the client organization?

What are the consequences of delaying the development of the system in terms of patient care?

What are the consequences of releasing a system that is potentially faulty?

Should you resign?

d) The client has asked you to check up on staff emails and, as the employer, they have a right to do this under the Regulation of Investigatory Powers Act. However, you are not part of the client organization and you should refuse to do this.

e) The BCS Code of Conduct, Section 4, states that: 'You shall ensure that within your professional field/s you have knowledge and understanding of relevant legislation, regulations and standards, and that you comply with such requirements. As examples, relevant legislation could, in the UK, include ... law relating to intellectual property.'

This code belongs to your former colleague and you should not use it directly. However, you can look at it for ideas and then write your own version, giving credit to your colleague in the documentation. On the other hand, people often sign contracts saying that the rights in anything they produce while employed belong to the employer. If this is the case, you would be able to use the code left by your former colleague.

f) The BCS Code of Conduct, Section 9, states that: 'You shall not misrepresent or withhold information on the performance of products, systems or services, or take advantage of the lack of relevant knowledge or inexperience of others.' You should be honest with the clients about what is happening on the project.

Chapter 11

Exercise 11.2
Points to be discussed could include the following.

- Symptoms. Systems are often late, over budget or not what was required.
- Possible causes. Today's systems are increasingly complex: we do not have enough properly trained software developers to satisfy demand.
- CASE tools solve many of the software developer's problems in using structured techniques: consistency checking, getting bogged down in detail, difficulty in modifying complex diagrams.
- CASE tools can help with documentation, version control and management of large projects.
- CASE tools support traditional methods. Is this the best way to develop the sorts of systems that are required today?

Exercise 11.5
RAD would be a suitable development approach for the jazz gigs and the high-fashion clothes systems only. The other examples are all too complex and safety or security-critical.

Exercise 11.8

Use case: Maintain product list

Actor: Administrator

Goal: To keep the product list up to date

Overview:

New products are added to the list; items no longer stocked are deleted; the product attributes can be edited.

Typical course of events:

Actor (Administrator) action	System response
1. Keys productNo of product	2. Displays product details to be deleted
3. Checks product details	
4. Selects delete option	5. Deletes product from database

Alternative courses:

Steps 1 to 5 For a new product the computer allocates a new productNo, the Administrator keys in the product details and the product is added to the database.

Steps 4 and 5 The Administrator can edit any of the product attributes.

Figure E11.8 Use case description of the Maintain product list use case

Exercise 11.9

- Sue receives an delivery of 50 packets of humorous birthdays cards from the supplier
- She keys in the product number
- The system displays the product details
- She increases the stock level by 50

Figure E11.9 Maintain product list scenario to update the stock levels

Exercise 11.10

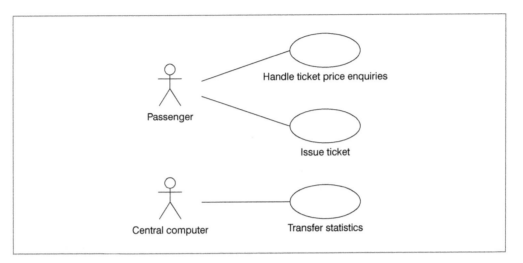

Figure E11.10 Use case diagram for the automatic ticket machine system

BIBLIOGRAPHY

Baase, S. (2003) *A Gift of Fire. Social, Legal and Ethical Issues for Computers and the Internet*, 2nd edn, Prentice-Hall, Upper Saddle River, NJ.

Begg, C. and Connolly, T. (2004) *Database Systems: A Practical Approach to Design, Implementation and Management*, 4th edn, Addison-Wesley, Harlow.

Bennett, S., McRobb, S. and Farmer, R. (2005) *Object-Oriented Systems Analysis and Design using UML*, 3rd edn, McGraw-Hill, London.

Britton, C. and Doake, J. (2000) *Object-Oriented Systems Development: A Gentle Introduction*, McGraw-Hill, London.

Britton, C. and Doake, J. (2004) *A Student Guide to Object-Oriented Development*, Elsevier Butterworth-Heinemann, Oxford.

Brooks, F.P. (1995) *The Mythical Man Month*, 2nd edn, Addison-Wesley, Reading, MA.

Date, C.J. (2003) *An Introduction to Database Systems*, 8th edn, Addison-Wesley, Reading, MA.

Fowler, M. with Scott, K. (1999) *UML Distilled: Applying the Standard Object Modeling Language*, 2nd edn, Addison-Wesley, Reading, MA.

Goodland, M. and Slater, C. (1995) *SSADM Version 4: A Practical Approach*, McGraw-Hill, London.

Hawryszkiewycz, I. (2001) *Introduction to Systems Analysis and Design*, 5th edn, Prentice-Hall, a division of Pearson Education, Australia.

Hoffer, J.A., George, J.F. and Valacich, J.S. (2004) *Modern Systems Analysis and Design*, 4th edn, Prentice-Hall, a division of Pearson Education, NJ.

Howe, D.R. (2001) *Data Analysis for Database Design*, 3rd edn, Butterworth-Heinemann, Oxford.

IEEE (2000) *0830–98 Software Requirements Specifications*, Institute of Electrical and Electronics Engineers, Inc., New York.

Kaposi, A.A. and Myers, M. (2001) *Systems for All*, Imperial College Press, London.

Kendall, E.K. and Kendall, J.E. (2005) *Systems Analysis and Design*, 6th edn, Prentice-Hall, a division of Pearson Education, Upper Saddle River, NJ.

Kotonya, G. and Sommerville, I. (1998) *Requirements Engineering: Process and Techniques*, John Wiley & Sons, Chichester.

Martin, J. (1991) *Rapid Application Development*. Macmillan, New York.

Myers, M. and Kaposi, A. (2004) *A First Systems Book: Technology and Management*, 2nd edn, Imperial College Press, London.

Pfleeger, S.L. (2001) *Software Engineering, Theory and Practice*, 2nd edn, Prentice-Hall, Upper Saddle River, NJ.

Pressman, R.S. (2004) *Software Engineering: A Practitioner's Approach*, 6th edn, McGraw-Hill Higher Education, London.

Robinson, B. and Prior, M. (1995) *Systems Analysis Techniques*, International Thomson Computer Press, London. (Unfortunately, not currently in print.)

Shneiderman, B. and Plaisant, C. (2004) *Designing the User Interface: Strategies for Effective Human–Computer Interaction*, 4th edn, Addison-Wesley, Reading, MA.

Skidmore, S. and Eva, M. (2004) *Introducing Systems Development*, Palgrave Macmillan, Basingstoke.

Sommerville, I. (2004) *Software Engineering*, 7th edn, Addison-Wesley, Wokingham.

Sommerville, I. and Sawyer, P. (1997) *Requirements Engineering: A Good Practice Guide*, John Wiley & Sons, Chichester.

GLOSSARY

Abstraction: the process of ignoring currently irrelevant details of a problem in order to concentrate on the most important parts.

Acceptance testing: the final testing of the software system in the presence of the client. The system is then accepted by the client or further alterations made.

Activity diagram: a diagram used in object-oriented development to model the details of complex processes.

Actor: in object-oriented development, a person, organization or physical device that interacts with the system in some way.

Agile Methods: a group of currently popular development methods, strongly influenced by prototyping and RAD. Two of the best-known Agile Methods are DSDM (Dynamic Systems Development Method) and XP (eXtreme Programming).

Analysis: analysis involves investigation into and modelling of both the problem area and the developing system.

Application independence: the separation of data storage from its applications.

Association: in object-oriented development, a link between two classes indicating a possible relationship between objects of the classes.

Attribute: data item or data element.

Automation boundary: delineates the part of the system that will be computerized.

Bar chart: type of graph. A technique used in project management where such diagrams (also known as Gantt charts) are used to help plan and schedule resources during the development of a system.

Beta testing: testing of a software product by typical users in a working environment before general release.

Candidate key: any attribute or combination of attributes that uniquely identifies an occurrence of an entity.

Capability Maturity Model (CMM): framework for assessing the level of maturity of a software company.

Cardinality: the cardinality of a relationship indicates the number of instances of each entity that are allowed to participate in the relationship.

CASE: Computer Aided Software Engineering. The use of software tools to automate the system development process. CASE tools can include any program that aids the system developer, such as a drawing program or a code debugger. However, the tools are usually much more sophisticated, covering several stages of the system life cycle.

Class: the description or pattern for a group of objects that have the same attributes, operations, associations and meaning. Template or factory for creating objects.

Class library: a collection of fully coded and tested classes that may be reused in other software applications.

Client: the person who requests the new system and is the main contact point with the system developer. The client will generally also be a user of the new system, but this is not always the case.

Code of conduct: guidelines produced by a professional body, such as the British Computer Society, which help software engineers with issues relating to ethical and professional behaviour.

Cohesion/cohesive: a module is cohesive if it has a clearly defined role, a single, obvious purpose in the application. This makes the module easier for a maintaining programmer to read and understand.

Collaboration diagram: in object-oriented development, illustrates the behaviour specified in a scenario, with the interactions organized around the objects and the links between them, rather than shown in a time sequence.

Configuration management: systematic logging of changes to the system; version control.

Consistency: a specification is consistent if there are no internal contradictions between different views of the system.

Context diagram: the top-level data flow diagram that is used to give an overall view of the current or required system.

CRC (class–responsibility–collaboration) cards: a technique in object-oriented development, using small index cards, to identify the responsibilities of classes in the system and the classes with which they have to collaborate to fulfil these responsibilities.

Critical path: in project management, identification of the critical path highlights those areas of the system development where any delay will cause a delay in the project delivery date.

Crow's foot: part of the notation for entity-relationship diagrams; indicates the 'many' part of a one-to-many or many-to-many relationship.

Current logical data flow diagram (CLDFD): this is a data flow diagram that is drawn to illustrate what happens in the current system without considering *how* it happens.

Current physical data flow diagram (CPDFD): this is a data flow diagram that illustrates how procedures are carried out in the current system.

Data dictionary: a structured modelling technique that uses text and a small set of symbols to define the data in the system.

Data encapsulation: see *encapsulation*.

Data flow diagram: a widely used structured modelling technique that models the movement of data round the system. Data flow diagrams are made up of processes, data flows, data stores and external entities.

Data hiding: see *information or data hiding*.

Data/information flow: one of the constituent symbols of a data flow diagram. The data flow is drawn as a directed line and is labelled with the name of the data.

Data model/modelling: the technique of building representations of a system based on the data objects or entities that are found in it. The principal technique used in data modelling is the entity-relationship diagram.

Data Protection Act: an Act of Parliament passed in the United Kingdom in 1984, amended in 1998, to protect personal data held in a form that can be automatically processed and that concerns identifiable individuals. The act stipulates that such data must be registered, protected against unauthorized access, and that individuals have the right to examine and correct the data concerning them.

Data redundancy: see *redundant*.

Data store: a symbol found on data flow diagrams. Data stores represent data that is stored permanently by the system, such as product prices or customer details.

Database: all the data required to support the operations of an organization collected, organized and maintained centrally and in such a way that it can be used by many different programs.

Database management system (DBMS): organizes the inputting, manipulation, retrieval and outputting of the data. The DBMS is the link between the way in which the data is stored and the way in which it is used by applications.

Decision table: a technique used in process definitions in which different conditions and actions are laid out in tabular form.

Decision tree: a technique used in process definitions in which different cases are represented in a tree structure.

Decomposition: the process of breaking a problem down into successively smaller parts in order to understand it better.

Degree of relationship: see *cardinality*.

Deliverable: the output from a stage in the system life cycle. Deliverables in the early stages of the life cycle are generally in the form of documents and diagrams. In the later stages they also include program code and test results.

Derived data: data that can be calculated from data that is already stored in the system.

Design: the stage in system development where the architecture of the system is determined: how the system as a whole is to be organized into smaller, more manageable components or sub-systems. Whereas analysis concentrates on understanding the user's view of the system, system design is driven by implementation concerns.

Determines: an entity A determines another entity B if each value of A is always associated with only one value of B.

Direct manipulation: a widely used interaction style, where the user moves objects on the screen corresponding to the tasks to be performed.

Documentation: the documentation of a system includes many different aspects – instructions for the users and operators about the running of the system; information necessary to those concerned with system maintenance, including the documents generated as deliverables during the system development.

Elicitation: see *requirements elicitation*.

Encapsulation: packaging data and operations into objects.

Enterprise rules: rules that govern the data in the system under consideration. Enterprise rules relate to the data model of a particular organization and define such things as the degree and optionality of relationships.

Entity: a unit or object of data that is part of the system under consideration and that is important in the development of the new system.

Entity life history: a structured modelling technique that illustrates how a data entity is affected by events over the course of time.

Entity-relationship (E-R) model/diagram: a structured modelling technique that identifies data objects in the system and illustrates the links between these objects.

Entity type: see *entity*.

Environment: the system environment refers to anything outside the system that affects it in some way – e.g. people or organizations generating or responding to system data.

Event: an instantaneous occurrence that is of significance to the system. An occurrence that triggers a state transition.

Event modelling: the modelling of ways in which the entities in a system behave during their lifetimes and how they respond to external events.

External entity: one of the symbols found on a data flow diagram. An external entity represents a person or organization that has links with the system but is not part of it.

Fagan inspection: a structured framework for investigating and uncovering errors in the output from any stage in the software development process.

Feasibility study: part of the system life cycle, which attempts to determine whether there is a practical solution to the problem under consideration.

Foreign key: if one table in a database contains an attribute that is the primary key of another table, this attribute is called a foreign key. A foreign key permits a link between the two tables.

Form: a form in a database such as Microsoft Access™ allows the user to display, enter and modify data on-screen. Forms are based on tables, but may include labels, boxes, lines and even pictures to make the user's task easier.

Framework: an agreed structure for developing software systems.

Functional decomposition: breaking down a system into smaller parts in terms of its processes.

Functional requirements: the tasks that a system is to perform, what its inputs and outputs are, and how these are linked.

Functionality: what a system does in terms of the processes it supports.

Gantt chart: see *bar chart*.

Hardware: equipment used by a system, including printers, keyboards and VDUs, disk and tape drives.

Implementation: the stage of the system life cycle where the design is translated into a programming language.

Implementation independent: an implementation-independent design can be implemented in different ways, using different programming techniques or languages.

Information or data hiding: making the internal details of a module inaccessible to other modules.

Information system: software systems that primarily store, retrieve and manipulate data, generally in a business environment.

Input: data that is entered into the system by the user.

Installation: the stage where the system is delivered and set up at the client's premises.

Instance: refers to (1) an individual occurrence of an entity, or (2) an object that belongs to a particular class.

Interaction diagram: diagram showing a set of messages that take place between objects to achieve a specific goal.

Interaction style: the way in which the system interacts with the user. Command line and direct manipulation are examples of interaction styles.

Internet: a vast network of networks that connect millions of computers throughout the world. The Internet supports the World Wide Web, electronic mail links and a huge number of specialized mailing lists.

Intersection entity: an entity that is introduced in an entity-relationship diagram to represent information stored about a relationship, rather than about an individual entity. An intersection entity breaks a many-to-many relationship into two one-to-many relationships.

ISO 9001: an international standard for quality assurance.

Life cycle: a recognizable pattern of steps taken to develop a software system. These generally include the key stages of requirements engineering, design, implementation and maintenance.

Logical model: the collected models (both text and diagrams) that illustrate what the system does, omitting consideration of how this is achieved.

Maintenance: the final stage of the life cycle, where errors are corrected and system modifications carried out.

Message: a request from one object to another that it execute one of its methods.

Method: a procedure that is part of an object.

Method/Methodology: recipe for developing a system. The detailed description of the steps and stages in system development, together with a specified list of inputs and outputs for each step.

Modelling: the process of building a representation of all or part of a system. This is carried out using techniques such as natural language, structured techniques, prototyping, formal specification languages or any combination of these.

Modification: changes of any size that are made to the system after it has been delivered to the client.

Multimedia: the combination of different forms of media – text, graphics, sound, photographs and video – in a computer-based system.

Multiplicity: see *cardinality*.

Network: the linking of two or more computers, thus allowing them to communicate, share data and applications and use the same peripherals.

Network chart: diagram showing task dependencies and identifying the critical path, used in project management.

Non-functional requirements: the attributes of a system as it performs its job. Non-functional requirements include usability, reliability and performance.

Normal form: the process of normalizing data involves following a series of steps going from the un-normalized form through first, second and third normal forms. Higher normal forms have been defined, but are not covered in this book.

Normalization: the process of organizing data items into groups in such a way that no redundant data items are stored. In this way unnecessary duplication of data is avoided.

Object: software packet containing data and operations on that data.

Object-orientation/object-oriented system development: an approach to developing software systems that is based on data items and the attributes and operations that define them.

Open source: open source is a new way of organizing the development of software and disseminating the source code. With open source software (OSS), program code is on the Internet and freely available to all. Users can modify the code, add to it and redistribute it as they please, subject to certain conditions – for example, that it cannot be sold and that it must continue to carry the open source conditions.

Operation: in object-oriented development, a procedure or function defined as part of a class or object.

Output: the information that is produced by the system for the user.

Pair programming: this involves two programmers working together, sharing the mouse, keyboard and monitor, collaborating on the same design, code and testing.

Physical model: concentrates on describing how a system works; physical models can be used to describe current and required systems.

Polymorphism: the ability to hide different implementations behind a common interface.

Portable: a portable program can be run on more than one type of computer.

Post-implementation testing: testing carried out after the system is coded (also known as code testing).

Pre-implementation testing: testing carried out before the system is coded.

Primary key: attribute, or a combination of attributes, that uniquely identifies an occurrence of an entity.

Problem domain: the area of knowledge or activity relating to the problem that the system is to solve.

Process/process modelling: in structured systems development a process is something that happens to data. A process is modelled in data flow diagrams by a box with data flows entering and leaving. What happens in a process is described in detail in the related process definition.

Process definition: a structured modelling technique that specifies the lowest-level processes on a data flow diagram.

Programmer: someone who transforms program or system specifications into a computer programming language.

Prototyping: an iterative method of developing a certain kind of system instead of using traditional structured methods. Disposable prototyping involves constructing a working model of the system at a very early stage in development and using the model to identify user requirements. Once these have been established, the prototype is thrown away. Evolutionary prototyping is where the working model is developed into the final system.

Pseudocode: a description, normally of a program or a process, that uses many of the features of a modern block structured programming language but has no rigid syntax. Features include the use of control structures (if ... then ... else, case, repeat) and a structured layout. Can be implemented in several different programming languages.

Quality assurance: a framework to support the production of high-quality software.

Query: an expression in a language such as SQL for retrieving and manipulating data. In commercial database packages queries are generated from forms that are filled in by the user.

Rapid Application Development (RAD): a popular development method based on prototyping, user participation and high-powered software tools.

Real-time system: a system that operates in real time, i.e. as dictated by the real world (e.g. patient monitoring or a missile guidance system).

Redundant: a stored data item is redundant if it is never used, if it is stored in more than one place in a system, or if it can be derived from other data stored in the system.

Relationship: a link between two entities, which is significant for the system.

Report: reports allow the user to format and display in a user-friendly form information that is output from the system.

Required logical data flow diagram (RLDFD): this is a data flow diagram that is drawn to illustrate what should happen in the new system without considering how it will be implemented.

Required physical data flow diagram (RPDFD): this is a data flow diagram that illustrates how procedures are to be carried out in the new system.

Requirements: a feature or behaviour of the system that is desired by one or more stakeholders.

Requirements elicitation: the stage of requirements engineering that aims to gather as much information as possible about the problem domain, the clients' and users' current difficulties and what they would like the intended system to do for them.

Requirements engineering: the process of establishing what is wanted and needed from a software system. Requirements engineering covers the three stages of requirements elicitation, specification and validation.

Requirements specification: the stage of requirements engineering during which the information from the requirements elicitation process is analysed and recorded using textual and diagrammatic modelling techniques to represent the problem and the proposed solution.

Requirements validation: the stage of requirements engineering which checks that the recorded requirements correspond to the intentions of the stakeholders about the system.

Responsibility: in object-oriented development, an obligation that one class has to provide a service for another.

Reuse: programming with existing software modules rather than coding them from scratch each time.

Safety-critical system: a system where failure would cause loss of life or severe hardship.

Scenario: in object-oriented development, a scenario represents one instance of a use case, describing a particular sequence of events that may occur in trying to reach the use case goal.

Security-critical system: a system where failure would cause loss of crucial information.

Self-transition: occurs when an object remains in the same state in response to an event.

Sequence diagram: in object-oriented development, illustrates the behaviour specified in a scenario, with the interactions shown in a time sequence.

Software: broad term covering the methods of using and controlling computers; includes programs and systems.

Software crisis: name given to the situation during the late 1960s and early 1970s, when software systems were typically late, over budget, unreliable, difficult to maintain and did not do what was required of them.

Software metrics: measurements that quantify the system development and the final software product.

SQL (Structured Query Language): a standard database query language.

Stakeholder: any person or organization affected by the system, such as users, clients, developers, management.

State: represents a period during which an entity satisfies some condition or waits for an event.

State diagram: diagram illustrating the behaviour of a single entity in response to events in the system.

State transition: the response of an entity to an event; usually involving movement of the entity from one state to another.

Structured approach: an ordered and clearly defined way of developing systems based on the traditional system life cycle that has been in use for several decades.

Structured English: a sub-set of English that is used in process definitions.

Structured walk-through: see *walk-through*.

Syntax: the rules that govern what constitutes a valid expression in a programming or modelling language.

System: a set of interrelated objects or elements that are viewed as a whole and designed by human beings to achieve a purpose; it has a boundary within which it lies and outside of which is the environment.

System boundary: defines what is to be considered inside the system and what will form its environment. On a data flow diagram the system boundary is defined on the highest level, context diagram.

System developer/designer: a general term for a member of the development team, often used to mean someone who does the work of an analyst, designer and programmer.

Systems analyst: usually a member of a data processing department. A systems analyst's work has traditionally been concerned with capturing user requirements, specifying and designing the new system. He or she will also be concerned with testing the system and maintaining it.

Table: repository of data in a relational database. Tables store data in a row–column format; each column stores a field, or attribute of the data, and each row stores a record, typically the complete set of values for a single data object.

Third-generation language: programming languages such as Pascal, FORTRAN, COBOL, which are oriented more towards the user and his or her problem than towards the machine, and are therefore considered to be more high level than the languages that preceded them.

Transaction processing system: a system that primarily stores, retrieves and manipulates data in a business environment.

Trialling: system tests done by the user once the system has been installed on site; usually, real past data is used.

Use case: in object-oriented development, specifies the functionality that the system will offer from the user's perspective. A use case specifies a set of interactions between a user and the system to achieve a particular goal.

User: any organization or person who uses the computer system to input or process data, or who receives the results of such processing.

User interface: the parts of the system with which the users come into contact, such as screens and reports.

Validation: the process of ensuring that what is being developed is actually what the client wants.

Verification: the process of ensuring that what is being developed is as free from error as possible.

Walk-through: a formal meeting of members of the development team to review a piece of work in detail.

World Wide Web: software on the Internet that allows users to access information held on computers worldwide.

Z specification language: a language based on maths and logic that is used to specify requirements for software systems, in particular safety-critical and security-critical systems.

INDEX